CHARGED!

BY

M.G. BUCHOLTZ, B.SC., MBA, M.SC.

A Wood Dragon Book

OTHER BOOKS BY M.G. BUCHOLTZ

FINANCIAL MARKETS

Stock Market Forecasting: *The McWhirter Method De-Mystified*
The Bull, The Bear, and the Planets: *Trading
the Financial Markets Using Astrology*
The Lost Science: *Esoteric Math and Astrology
Techniques for the Market Trader*
The Cosmic Clock: *Timing the Financial Markets Using Astrology*
Financial Astrology Almanac 2024: *Trading &
Investing Using the Planets (11th edition)*
Follow The Trend: *When to Buy and When to Sell*

SCIENCE

The Recipe: *Reviving the Lost Art of Home Distilling*
Field to Flask: *The Fundamentals of Small Batch Distilling (5th Edition)*
Frozen Fury: *Agricultural Crops and Hail Damage*

POLITICS AND SOCIAL ISSUES

Thatcher versus Douglas: *The CCF, the Liberals,
and the Mossbank Debate of 1957*

CHARGED!

THE DANGEROUS AND MISGUIDED PROMISE OF THE ELECTRIC VEHICLE

BY

M.G. BUCHOLTZ, B.SC., MBA, M.SC.

A Wood Dragon Book

CHARGED!
THE DANGEROUS AND MISGUIDED PROMISE
OF THE ELECTRIC VEHICLE

Published by:
Wood Dragon Books
Box 429, Mossbank, Saskatchewan, Canada, S0H 3G0
http://www.wooddragonbooks.com

Cover design by: Callum Jagger/Hyperlight
Artwork Inside Design by: Christine Lee

ISBN: 978-1-990-863-67-7 (Paperback)
ISBN: 978-1-990-863-68-4 (Hardcover)
ISBN : 978-1-990-863-66-0 (eBook)

To contact the author: supercyclereport@gmail.com

CONTENTS

DEDICATION

To all who sense that the push towards electric
vehicles is misguided, this book is for you.

Read it, learn from it, and spread the word:
electric vehicles are *not* going to save our planet.

INTRODUCTION

As a young engineering student at Queen's University in the early 1980s, I held my professors in high regard and viewed them through a lens tinged with fear. These were smart men, educated at world-renowned institutions in Europe and North America. It was daunting for me, a recent high school graduate, to sit in a lecture hall listening to them describe the principles of chemistry, thermodynamics, and materials science.

One of my professors left a lasting impression on me thanks to a sharp rebuke he levelled at our entire class. He told us very bluntly that we were *not* at Queen's University to become engineers. We were there to learn how to think and to learn how to solve problems. This stern reminder has stayed with me ever since and in any problem-solving task I approach, I first aim to identify and understand the root cause. I avoid the temptation to just jump to a final answer.

This mindset greatly assisted me with my MBA degree studies in the late 1990s. My professors at Heriot-Watt University in Edinburgh, Scotland warned us that simply reciting the material from the textbooks on a final exam essay would *not* earn a passing grade. We would be required to demonstrate a deeper understanding of the principles of finance beyond what the textbooks described. To gain this deeper understanding, I devoured back-issues of the *Economist* and *Fortune 500*. Because of these rigorous demands, I came to hold my professors at Heriot-Watt in high regard, much as I had done years earlier at Queen's University.

In 2017, I returned to Heriot-Watt University to pursue my M.Sc. degree. At the age of 54, I was by far the oldest student in the class; I was even older than most of my professors. From the start of my M.Sc. studies, I sensed that something was different. There were no urgings about thinking, problem solving, or going beyond the textbook. I sensed that education had changed. It had become less rigorous.

British educational requirements mandated that my course load include one elective course. After looking at the list of available options, I settled on *Renewable Energy*. I was curious to learn more about solar panels and wind turbines. This course certainly provided detailed technical knowledge, but the course material shocked me in an unexpected way. During one particular lecture, the professor introduced us to the most recent *Intergovernmental Panel on Climate Change* (IPCC) report. He was upset by the report. He emotionally told the class that unless mankind took more assertive action to thwart climate change, the world as we knew it would cease to exist by 2030. At first, I thought he was joking. He was not. He fervently believed everything contained in the IPCC reports. He was a faithful, unwavering servant of the IPCC and the United Nations. He questioned none of their data, none of their conclusions.

Every time I read about a so-called expert expressing the urgency for buying an electric car, or hear a politician waxing eloquently about zero-carbon emissions, I think back to my Heriot-Watt experience. We are losing our ability to think and to solve problems. We are reaching for what we think is the final answer. We are ignoring the root causes.

Who are the wizards behind the curtain promoting this incessant drive for green energy, zero-carbon emissions, and electric vehicles? How have they so profoundly influenced politicians, bureaucrats, and academics? Once I start to question an idea, policy, or platform, deep in my creative headspace a book begins to take shape.

Adopting electric vehicles is a theme that our political leaders and bureaucrats have not thoroughly thought through. Our politicians, with support from the media, have jumped to the final answer that climate change is an anthropogenic (man-made) emissions phenomenon and that it can be remedied merely through changes to human behavior. The band-aid remedy is for the world to adopt electric vehicles.

In my *Renewable Energy* course, we were expected to blindly accept that the world as we knew it was in danger of ending. We were expected to recite IPCC conclusions on our final exam answers. If this experience is indicative of education at large, I argue that society is losing its ability to critically think and is in danger of just accepting the final answer to problems as dictated to us by politicians, academics, and media.

Instead of marching off to a car dealership to look at the available models of electric cars, people should first be doing their homework. How does the battery in a typical electric vehicle work? How is it constructed? How does it differ from the batteries in gasoline-powered vehicles? Is a wholesale shift to electric vehicles even feasible? Is it sustainable? Throughout automotive history, has the concept of the electric car been tried before? What was the outcome? What are the geopolitics that pose a threat to the electric car theme? Where does the electricity come from to recharge an electric vehicle battery? What infrastructure development plans are in place to ensure electric vehicle users will be able to recharge their batteries in a timely fashion whenever needed?

This book explores the backstory behind the push for electric vehicles, starting with a look at the history of the energy sources used to power vehicles, and the history of energy storage using batteries. A brief look at the history of metal hydride (NiMH) and lithium-ion (Li-ion) batteries then leads to an exploration of the electrochemistry of lithium-ion batteries, and a critical examination of the availability and sustainability of the metals, minerals, and chemicals used in making lithium-ion batteries. The question of where electricity comes from is then explored in the context of the immutable Laws of Thermodynamics.

Woven into this book is an explanation of the root causes of climate change. The Earth exhibits long cycles of variable eccentricity, tilt, and spin in its orbit around the Sun. These long cycles play a significant role in the variability of our climate and weather patterns. These long cycles cannot be changed. The eight billion people on the planet are also weighing on the ability of Mother Nature to carry us all. If we are to enjoy a good quality of life going forward, collectively we need to reduce energy use, reduce emissions, and pare back on our consumption. This conclusion, among others expressed in this book, is at extreme variance to the message being touted by academics, politicians, the media, and elitist groups like the IPCC, the World Economic Forum, and the Club of Rome. Blindly grasping onto a policy that favors electric vehicles is a foolish approach to helping the planet. Policy makers, academics, elected officials, and indeed individual consumers need to all stop and assess the very meaning of our existence on the planet. The electric vehicle theme is a fear-based, knee-jerk reaction to a very complex problem.

In late 2023 and into early 2024, as I was finishing the final edits on this manuscript, I could sense an awakening of sorts towards the electric vehicle theme. The following news releases caught my attention:

October 13, 2023 (Reuters) - A United Autoworkers (UAW) official said in a memo that Ford was considering canceling a production shift for electric vehicle production due to slowing demand and that the company was looking to build more gasoline-powered trucks instead. *It doesn't take a rocket scientist to figure out that our sales for the F-150 Lightning have tanked,* the memo read.

October 17, 2023 (Bloomberg) - The Board of Directors of Swedish electric truck maker Volta Trucks announced that uncertainty with their battery supplier has negatively affected their ability to raise sufficient capital in an already challenging capital-raising environment for electric vehicle players. With deep and sincere regret, the Board has made the difficult decision to take steps for Volta Trucks to file for bankruptcy proceedings in Sweden.

November 21, 2023 (Reuters) - Ford is "re-timing and resizing some investments" through scaling back plans by nearly $1.5-billion at its EV battery plant in Michigan as it adjusts to market demand. The company is cutting production capacity at the facility by over 40%. Ford now expects the facility to produce around 20 GWh, a big difference from the 35 GWh initially expected. Ford is cutting the expected jobs to 1,700 from an initial planned 2,500.

January 11, 2024 (Reuters) - Rental firm Hertz Global Holdings (HTZ.O) is selling about 20,000 electric vehicles, including Teslas, from its U.S. fleet two years after a deal with the automaker to offer its vehicles for rent, in another sign that EV demand has cooled. Even though it had aimed to convert 25% of its fleet to electric by 2024 end,

Hertz will instead opt for gas-powered vehicles, citing higher expenses related to collision and damage for EVs.

January 26, 2024 (Fortune Magazine) - Tesla shares were down just over 12% on Thursday after CEO Elon Musk sounded the alarm over Chinese electric carmakers which he called the "the most competitive car companies in the world." Musk claimed that Chinese carmakers (which include BYD, Geely, and SAIC) are "extremely good" and could threaten other carmakers in the U.S. and elsewhere if governments do not step in. "Frankly, if there are not trade barriers established, they will pretty much demolish most other car companies in the world," he said.

These news releases perfectly set the stage for the arguments to be made in this book. There are many reasons a continued shift towards a society mobilized by battery-powered vehicles is not possible. The information you are about to read will show that pursuing the promise that electric vehicles will save the planet is not only unsustainable, it is misguided and dangerous.

SYMBOLS AND UNITS GLOSSARY

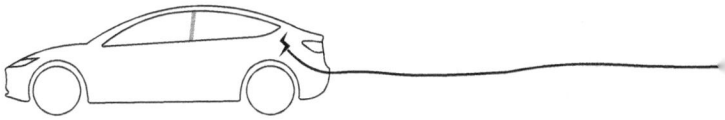

This book uses various technical terms and units of measure to explain the science of batteries. The following are the technical terms and various units of measure referred to in this book, along with a brief explanation of their origin.

Horsepower (Hp): The horsepower unit of measure is defined as the work needed to raise 550 pounds of mass through 1 foot of height in 1 second. Scottish Engineer James Watt defined the horsepower unit of measure in the late 1700s to compare the output of steam engines with the power of draft horses. One horsepower is equal to 746 Watts.

Watt (W): The Watt is named in honor of James Watt. One Watt is equivalent to 1 Joule per second. In electricity calculations, volts multiplied by amps equals Watts.

kiloWatt hour (kW·hr): The kW·hr is the typical unit of consumption for electrical energy. For example, if a house was equipped with an electrical circuit providing 40 amps of energy at 240 volts, a homeowner plugging a device into this circuit for 1 hour would consume 9.6 kW·hr of energy. At a cost of, say, 14 cents per kW·hr, the homeowner would incur a bill from the utility provider for $1.34 for using this quantity of power.

Capacity (A·hr): The capacity of a battery refers to the amount of energy it can store. The unit of measure for capacity is the Amp-hour (A·hr). An individual cell, many thousands of which comprise the battery pack in an electric vehicle, will have a capacity of around 3 A·hrs.

Total Energy (W·hr): Individual cells in an electric vehicle battery have a rated amount of total energy that can be stored. If an individual cell with a capacity of 3 A·hrs produces 3.7 volts, the total energy of the cell is 11.1 W·hrs or 0.0111 kW·hrs. Assembling, say, 7000 of these individual cells into a battery pack will produce a battery with a total energy of 77.7 kW·hrs.

Energy Density (Watt-hours): The energy density (also called the specific energy density) of a battery refers to the energy stored in the battery per unit of mass or volume. The unit of measurement for energy density is either Watt-hours per kg or Watt-hours per liter. A typical lead-acid battery has an energy density of about 40 Watt-hours per kg. A lithium-ion battery has an energy density of about 200 Watt-hours per kg.

Gigawatts (GW): Battery manufacturing plants often express their production output capacity in Gigawatts, or more precisely Gigawatt hours (GW·hrs). A battery manufacturing plant producing 257,400 battery packs per year each rated at 77.7 kW·hrs (77.7 x 257,400/1,000,000) would be termed a 20 GW plant.

Newton (N): The Newton is named in honor of Isaac Newton. One Newton is the force required to move a 1 kg mass from a standstill to a velocity of 1 meter/second in a time of one second.

Joule (J): One Joule is equal to a force of 1 Newton acting to move an object through a distance of 1 meter. One Joule is 1 N·m or 1 kg·m^2/sec^2. In terms of electrical energy, one Joule is the work required to produce one Watt of power for one second (1 Joule = 1 Watt-second).

Discharge Rate (amp): The discharge rate of a battery is the maximum current that a battery can provide to an electric device, such as the motor that drives the wheels on an electric car. The *maximum peak discharge rate* refers to the amount of current the battery can provide over a short burst of time. The *maximum continuous discharge rate* refers to the maximum amount of current the battery can deliver over an extended time without overheating. The unit of measurement for discharge rate is the amp.

C Rate: The C rate of a battery refers to the rate of discharge or the rate of charging that a battery is capable of. It is calculated by taking the discharge rate and dividing by the capacity. For example, a battery with a 4 amp maximum continuous discharge rate and a 2 amp-hour capacity will have a C Rate of 2. A battery with a 4 amp-hour capacity under a 2 amp discharge rate will have a C Rate of 0.5. A lower C Rate means the battery is capable of more charge/discharge cycles and will have a longer life.

Maximum Cycles: Maximum cycles is a reference to the number of charge/discharge cycles that a battery can endure where each recharge takes the battery to 80% or more of its original capacity.

Fuel Efficiency (km/L): This measure of fuel efficiency describes how many kilometers of distance can be driven on one liter of fuel. One km/L is equal to 2.352 miles per U.S. gallon (mpg).

Mole (mol): a quantity of measure calculated as mass/molar mass, where molar mass is obtained by referencing the Periodic Table of the Elements.

MANKIND LEARNS TO STORE ENERGY

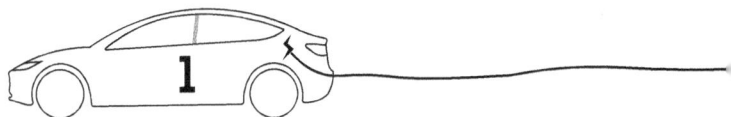

LEYDEN JARS

The history of storing energy in a battery goes back to the mid-1700s.

Thinkers and tinkerers at that time became fascinated with the idea of capturing electricity. German professor Georg Bose generated static electricity and then used a spark of this electricity to ignite an alcohol-water mixture in a jar. News of this experiment soon reverberated across Europe.

German cleric and inventor Ewald von Kleist took Bose's experiment one step further. He filled a medicine bottle with a water-alcohol mixture. A nail was inserted through the cork of the bottle. When the nail was hooked to a static electricity generating device, the static electricity flowed through a wire connected to the nail and then into the water-alcohol mixture causing a buildup of electric charge.

Meanwhile in Leiden, Netherlands, scientist Pieter van Musschenbroek was also busy replicating Bose's experimental work. He called his bottle device a *Leyden Jar*. News of the Leyden Jar soon spread to the colonies in America.

One of the images I recall from grade school is that of Benjamin Franklin flying a kite in a lightning storm so he could better understand electricity. We were told that Franklin was a talented inventor who understood electricity. Of course, we were never told the whole story. In fact, Benjamin Franklin decided to make his own design of a Leyden Jar, minus the actual jar. He took 11 panes of glass with thin lead plates glued to each pane and then sandwiched the panes together. He then assembled a kite with a long string made of hemp. He attached a metal key to the hemp string. The string was then connected to the sandwiched panes. As he flew the kite in a storm, the static electricity from the atmosphere accumulated on the metal key, traveled through the wet hemp string, and collected in the sandwiched Leyden plate device. He is later credited with using the term "electrical battery" to describe his kite experiment. Other inventors, tinkerers, and professors went on to refine the capture of electricity. Mankind was learning how to harness the power of Nature.

VOLTAIC PILES

In the 1780s, the science of electricity took a significant leap forward thanks to frog legs. Luigi Galvani was an Italian physicist and anatomist with a particular fascination for dissecting frogs. When he mounted a frog leg on a metal hook made of brass and proceeded to probe the leg with an instrument made of a different metal, the leg could be seen to twitch. Galvani was convinced he had discovered a new form of electricity. Italian researcher Alessandro Volta took issue with Galvani's theory, arguing that the use of two different metals was merely creating a weak current which caused the muscle tissue in the frog leg to contract. To illustrate his point, Volta created an experiment in which he assembled a stack of alternating zinc and copper discs separated by bits of cloth that had been soaked in a salty solution. A wire was connected to each end of the stack of discs. Inserting the wire ends into a container of water produced tiny bubbles. The stacked device was generating a flow of electricity that was causing the hydrogen and oxygen atoms in the water molecules to separate. Volta's device became known as the *voltaic pile*. Although not realized at the time, this device would go on to define what we today call a *battery*.

DANIELL CELLS

Building on the science of two different materials generating a flow of electricity, English chemist John Frederic Daniell created the *Daniell cell* in 1836. His device comprised two containers, one with a copper electrode and one with a zinc electrode. The containers were filled with copper sulfate solution and zinc sulfate solution respectively. The two containers were joined by a U-shaped tube that was filled with a salt solution. To complete the circuit, the two electrodes were joined by a piece of copper wire. Daniell showed that electrons from the zinc electrode would flow through the copper wire to the copper electrode. Daniell called the zinc electrode the *anode*. The copper electrode was called the *cathode*. Positively charged zinc ions would flow through the U-shaped tube towards the copper cathode. Negatively charged sulfate ions would flow through the U-shaped tube towards the zinc anode to balance the circuit. Daniell further demonstrated that the reaction would continue so long as zinc electrode material remained. Daniell's cell eventually came to be called a *voltaic cell*, named in honor of Alessandro Volta. Several years later, in 1843, English inventor Robert Bunsen created the *Bunsen cell*. His device comprised a coiled sheet of zinc as the anode in dilute sulfuric acid and a carbon cathode in a bath of nitric acid. The two baths were separated by a porous ceramic layer inspired by the ceramic clay material that his tobacco smoking pipe was made from. Where Bunsen's cell improved upon Daniell's design was through its compact, containerized design.

PLANTE AND JUNGER

The next significant advance in electricity science came nearly two decades later thanks to the work of French scientist Gaston Plante. After graduating from Sorbonne University in 1855, Plante set up a laboratory in Paris to further study the electricity phenomenon. In 1860, he unveiled his lead-acid energy storage system. Over the next dozen years, his lead-acid energy storage device came to be used to power items like miners' electric lamps, electric bells, signal horns, lights on ships, and electric brakes on steam trains.

Nearly 40 years later in 1899, Swedish inventor Waldemar Junger offered what he deemed to be an improvement over Plante's lead-acid design. His device used nickel as its positive anode, and cadmium as its negative cathode. While his design could be charged quicker than Plante's lead-acid device, the nickel and cadmium materials made the cost of his design too high.

THOMAS EDISON

Thomas Edison of the Edison Power Company was watching Junger and his nickel-cadmium design. When it became apparent that Junger's design would be too costly to produce, Edison drew attention to himself in 1900 when he revealed a less costly nickel-iron design. The cathode was made of nickel oxide material. The anode was made of iron. The electrolyte was a solution of potassium hydroxide. His design could be charged quicker and had a significantly higher energy density than Plante's lead–acid batteries. After obtaining patents for his nickel-iron design, he set up a production facility, Edison Alkaline Storage Batteries, to further his dream of seeing electric cars powered by his nickel-iron battery design. But it would not come to be. Vehicles powered by Edison's battery could not surpass the internal combustion engine which was aided by abundant oil from Texas, the ability to refine that oil into combustible gasoline, Kettering's starter device which drew current from a lead-acid battery, and Henry Ford's affordable vehicle prices. Moreover, Edison's batteries had two inescapable flaws that further dimmed the prospects for electric cars: his batteries performed poorly at low temperatures and were more expensive than Plante's lead-acid design.

THE MODERN LEAD-ACID BATTERY

If you drive a vehicle powered by an internal combustion engine, it will be equipped with a modern version of Plante's lead-acid storage device. We take this device for granted and refer to it as a *battery*. The energy contained in a lead-acid battery provides electricity to the electric starter and to the engine's spark plugs. The spark plugs initiate combustion of a compressed fuel-air mixture in the engine's cylinders. The combustion reaction drives the pistons in the cylinders downwards and causes a crankshaft to turn. The crankshaft is connected to the drive shaft which is connected to either the rear wheels or the front wheels. As the wheels turn, the vehicle moves. Part of the kinetic energy generated by the engine is used to turn a device called an alternator which generates a current that keeps the lead-acid battery charged.

A typical automotive lead-acid battery consists of six cells. Each cell comprises a positive electrode (cathode) and a negative electrode (anode). The cathode is made of lead dioxide. The anode is made of a porous lead material called *sponge lead*. The electrodes are immersed in a solution of battery acid (dilute sulfuric acid). This electrode assembly is contained in a plastic structural case which rests under the hood of the vehicle. The chemical reaction that occurs at the anode (negative electrode) involves the sponge lead dissociating into a lead ion and two electrons. The lead ion then reacts with SO_4^{2-} ions from the dilute sulfuric acid:

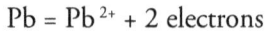

$$Pb = Pb^{2+} + 2 \text{ electrons}$$
$$Pb^{2+} + SO_4^{2-} = PbSO_4$$

The two electrons migrate to the cathode. The chemical reaction that occurs at the cathode (positive electrode) involves lead dioxide reacting with the protons from the acid and the gained electrons:

$$PbO_2 + 4\,H^+ + 2 \text{ electrons} = Pb^{2+} + 2\,H_2O$$

The overall battery cell reaction is:

$$Pb + PbO_2 + 4H^+ + 2\,H_2SO_4^- = 2PbSO_4 + 2H_2O$$

The voltage provided by a typical lead-acid battery cell is 2 volts. This explains why a vehicle battery with six cells will deliver 12 volts of energy. The production of H_2O molecules over time dilutes the acid solution which gradually impairs the efficiency of these reactions. This explains why, over time, a typical vehicle battery will run out of power and not be able to start your vehicle. A typical lead-acid battery is good for about 1,500 vehicle starts, unless those starting efforts are made in a cold weather climate when the engine is cold.

When the mechanic at a local garage or car dealership replaces the battery, the spent battery is sent away for recycling of the lead components and acid solution. The reason lead-acid batteries remain environmentally acceptable is the recyclability of their internal lead components. The U.S. Environmental Protection Agency (EPA) claims that close to 99% of all lead-acid batteries are recycled.

Unfortunately, all the engineering efficiencies and recycling capabilities of the lead-acid battery are in danger of being shoved aside by the lithium-ion battery movement. Our political leaders have decided that vehicles with internal combustion engines do not fit their vision of a zero-emission society.

- The late 1700s was a period of fascination with electrical energy. Work by inventors such as Georg Bose in Germany, Pieter van Musschenbroek in Holland, and Benjamin Franklin in America showed that electrical energy could be captured from Nature.
- Luigi Galvani dissecting frog's legs with instruments made of different metals and Alessandro Volta's criticism of Galvani's work culminated in the knowledge that two different metals connected by a conductive liquid could create electricity—a flow of electrons.
- In the 1830s, English chemist John Frederic Daniell refined Alessandro's work using a copper-zinc pairing of metals. English inventor Robert Bunsen crafted a more compact zinc-carbon cell. Mankind was now on its way to creating and storing electrical energy.
- In 1860, French chemist Gaston Plante unveiled his lead-acid cell which went on to become the basis for the 12-volt batteries used in vehicles today.

- In 1900, Thomas Edison unveiled his nickel oxide-iron cell design. This type of battery is still used to this day in industrial machinery applications.
- Despite all the work done to create affordable battery designs that can easily be recycled, political leaders and global thought leaders have now decided that vehicles with internal combustion engines do not fit their vision of a zero-carbon society and electric cars powered by expensive lithium-ion batteries will define the way forward.

ELECTRIC VEHICLES LOSE THE FIRST MOBILITY RACE

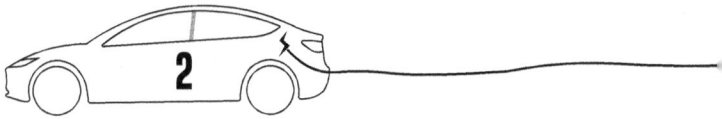

The late 1800s saw a significant technological advance in human mobility. Prior to this advance, people moved on foot, on trains powered by steam, or by horse-drawn carriage. The technological shift that changed the way people move was not a smooth one; it was a veritable battleground for dominance. The battle pitted electricity against hydrocarbon fuels like gasoline and kerosene.

THE GASOLINE-POWERED INTERNAL COMBUSTION ENGINE

The gasoline-powered, internal combustion engine traces its beginnings to Germany. In 1864, German inventor Nicolaus Otto patented a design for an internal combustion engine in which the ignition of a compressed fuel-air mixture was initiated by a timed spark. In 1876, Otto began working with inventors Gottlieb Daimler and Wilhelm Maybach to patent the compressed-charge, four-cycle engine. In 1886, German inventor Karl Benz introduced his Benz Motorwagen three-wheeled vehicle, powered by a four-stroke internal combustion engine. This design had a top speed of nearly 10 miles per hour but was difficult to steer and control. In 1888, following several design improvements, the Motorwagen Model 3 began to sell commercially in Germany and France.

In the U.S., in 1893, brothers Charles and James Duryea introduced a one-cylinder, four-horsepower design of vehicle called the Duryea Motor Wagon. Capable of traveling seven miles per hour, this was the first gasoline powered vehicle offered for sale in the U.S. The brothers parted ways several years later and brother Charles set up his own shop with a group of former employees to make a three-wheeled vehicle powered by a three-cylinder gasoline engine, capable of a 20 mile per hour top speed. Charles Duryea would soon face some stiff competition.

By 1900, there were near 100 brands of automobile being made across the U.S. bearing brand names like Hupmobile, Pierce Arrow, Franklin, Wolseley, Albion, Cleveland, and St. Louis. In 1901, Ransom Olds was busy producing his Curved Dash vehicle powered by a gasoline engine. Olds decided to bring efficiency to his production efforts. He engaged subcontractors to build the various vehicle components. He then set up an assembly line to efficiently piece together the components. The names of some of these subcontractors would one day be associated with well-known automobile brand names. The Dodge brothers supplied transmissions to the Olds Motor Company. Engines were supplied by Henry Leland; the Cadillac and Lincoln brand names would later be created by Leland. Fred J. Fisher who supplied vehicle bodies to Olds would eventually rise to prominence as the body maker for General Motors.

STEAM POWER CREATED BY BURNING KEROSENE

The first several years of the 1900s saw steam-powered vehicles with a commanding 40% of the vehicle market share. The steam to empower the pistons in the engine was created by burning kerosene. The thermal energy from combustion of the kerosene fuel was used to heat water in a tank to the boiling point. Gasoline-powered vehicles with internal combustion engines were behind, but gaining fast with 22% of the market share. The leaders in the steam-powered design field were Maine schoolteachers Francis and Freeland Stanley with their Stanley Steamer model. The brothers got into the car business using the money from the sale of their photographic plate business to camera company Eastman Kodak. The Steamer was powerful and capable of speeds well in excess of internal combustion engine cars. However, the vehicle operator had several challenges including having to wait for the boiler to heat up, and having to master the operation of levers, valves, and pressure gauges to make the vehicle move.

ELECTRIC POWER

In the early 1900s, electric vehicles commanded around 38% of the vehicle market. The popularity of electric vehicles traces its start to France in 1873 when inventor Zenobe Gramme created the first direct current (DC) generating device. The device could be used to add charge to a lead-acid electrochemical cell (a battery). At the Paris Exposition Internationale d'E'lectricite in 1881, electrician Gustave Trouve displayed an electric tricycle powered by a lead-acid battery. Shortly afterwards, inventor Paul Pouchain introduced an electric vehicle capable of carrying six passengers while moving at 16 km per hour. In 1898, engineer Charles Jeantaud unveiled his Jeantaud Duc model of electric vehicle powered by a 26.8 kW·hr (35.9 horsepower) battery. Jeantaud's design set the first official land speed record for

CHARGED!

electric cars, reaching a speed of 63 km per hour. Not to be outdone, several months later Belgian inventor Camille Jenatzy guided his electric Jamais Contente design to a new record speed of 105 km per hour. Even though the electric vehicle was drawing attention and setting speed records it had two significant practical limitations: its battery was heavy, and the driving range on a fully charged battery was only about 32 kilometers.

The popularity of the electric vehicle in Europe did not go unnoticed in the U.S. In 1890, William Morrison unveiled a four-wheeled design of vehicle, powered by lead-acid batteries and capable of transporting six passengers at a speed of 13 miles per hour. The popularity of electric vehicles in the U.S. was aided in large measure by Thomas Edison and his growing network of direct current (DC) electric power lines. Thanks to Edison, the technology to support battery recharging stations was readily available. By 1897, electric taxi cabs were a common sight in New York City and Philadelphia. A number of electric car companies was soon established including Pope Manufacturing, Anthony Electric, Baker, Detroit, Edison, Studebaker, Columbia, Anderson, Bailey, Chapman, Rausch & Lang, Waverly, and Woods. To lend further support to the growing industry, Thomas Edison developed the alkaline nickel–iron battery. This design of battery had higher energy density than a lead-acid battery and could be charged in half the time.

One of Edison's star employees was a Michigan farm boy with tremendous mechanical aptitude. His name–Henry Ford. On evenings and on weekends when not working for Edison, young Henry immersed himself in his fascination for internal combustion engines. He also had a keen understanding of how people like Ransom Olds were using efficient production methods to keep vehicle manufacturing costs down. Edison tried to convince young Henry that there was no future in cars with engines that ran on kerosene and other equally foul-smelling fuels. As Edison was making his arguments to young Henry Ford, something else was developing behind the scenes that would soon disrupt the vehicle market.

ROCK OIL

In the late 1800s, western Pennsylvania was viewed by scientists as a geological anomaly. In several remote areas of the state, a pungent fluid could often be seen oozing to the surface. People called it *rock oil* or *Seneca oil* in honor of a local Indian chief who had taught his people to apply it to their bodies to cure everything from headaches to upset stomachs.

A former school teacher turned lawyer, George Bissell, became aware of rock oil and in 1853 decided to commercialize it as a fuel for lamps. For years, people had been using whale oil or coal oil to keep their lanterns illuminated. Bissell reasoned that if he could collect rock oil cheap enough, he could sell it at a lower price point than whale oil or coal oil and take a big chunk of the lantern fuel market. Bissell and his investors hired Benjamin Silliman, a Yale University chemistry professor, to determine if rock oil would be a suitable rival to coal oil and whale oil.

In 1859, Bissell and his investors drilled a 69-foot hole into the ground in western Pennsylvania. The drill bit hit rock oil and more holes were soon drilled. As liquid oozed from the drill holes, Bissell and his group scrambled to find enough barrels to store their collected rock oil. But the scramble for storage was premature; the rock oil did not flow indefinitely. Over the next several years, as more and more entrepreneurs drilled for oil and as various wells ran dry, the price of rock oil went on wild, up and down roller-coaster excursions.

Despite the volatile price of rock oil, the area around the nearby industrial city of Cleveland, Ohio became the focal point for distillery operations to process the rock oil from the various wells drilled in western Pennsylvania. The distilleries would heat the rock oil to separate off the higher boiling point impurities. This new industry caught the attention of John D. Rockefeller, an entrepreneur involved in buying and selling commodities such as wheat, salt, and pork bellies. He aggressively became involved in the rock oil distilling business. Rockefeller made so much money in the distilling business that he funded the creation of a new university known today as the University of Chicago.

Chemists, like Silliman, who were studying the emerging rock oil phenomenon determined that the first vapors that came off the distillation process did so at just over 40°C. This material, called *kerosene*, comprised shorter molecules with 5 or 6 carbon atoms and became highly desirable as a fuel to light lamps. Material with longer molecular chains that came off the distillation process starting at about 100°C earned the name *gasoline*. With no ready use for it, the gasoline was often discarded. That is, until it was discovered that gasoline was very efficient at combusting in internal-combustion engines.

TEXAS BOOMS

Despite the technical advances made in distilling, the rock oil industry was still faced with the possibility of wells suddenly going dry. Word began to circulate that entrepreneurs were finding oil occurrences in California and sporadically in Texas. In Beaumont, Texas, entrepreneur Patillo Higgins noticed an elevated, domed piece of land on the outskirts of town. He visited some deep-pocketed investors in Pittsburgh and described why he felt the domed land feature was a prime location to drill for rock oil. They took him up on his quest and on January 10,

1901, the drill rig hit rock oil. Before long Higgins and his investors had drilled a number of wells that collectively were producing 75,000 barrels per day. The Texas rock oil boom was on in full force. Gasoline was about to become a readily available commodity. Rock oil earned a new name–*crude oil*.

HENRY FORD TAKES NOTE

Henry Ford took note of the developments in Texas and bid his employer, Thomas Edison, farewell. Ford took Ransom Olds's idea of subcontracted vehicle parts to a new level. He set up an assembly line in Detroit in such a way that an individual worker was responsible for repetitively performing a particular task. The time to construct a vehicle shrank from hours to minutes and the cost of producing a vehicle dropped to half that of an electric car. The average person could now afford one of Henry Ford's Model T cars priced at $650. The gasoline-powered vehicle was now poised to dominate the automobile market.

NO MORE CRANKING

However, gasoline-powered, internal-combustion engine vehicles had one thing in common. They had to be started through the action of a manual crank. Cranking a vehicle to start it was physically demanding and potentially dangerous.

This would all change in 1912 thanks to an inventor named Charles Kettering. After graduating from Ohio State University with a degree in electrical engineering, he was hired by the National Cash Register (NCR) company. He implemented a number of improvements to electric cash registers, but he was enamored with the emerging automobile market. On weekends he got together with some of his NCR workmates to tinker on electrical improvement idea for cars. One idea they worked on was a device that could draw power from a lead-acid battery and generate a high-voltage spark to the cylinders of a gasoline-powered engine. They envisioned replacing the manual cranking action to start vehicles with the simple press of a button located inside the vehicle. Their vision came to fruition and the electric starter was developed. In 1912, Kettering's efforts were noticed by car maker Henry Leland who ordered 12,000 starter motor units for his Cadillac brand of cars. Other auto makers soon placed orders too. Thanks to Kettering, the need to crank-start a vehicle faded away. The difficult-to-operate steam car also faded into the annals of history. The electric car with its heavy battery and limited driving range disappeared–aided in large part by cheap oil from Texas, Henry Ford's affordable cars, and Charles Kettering's electric starter.

The gasoline-powered vehicle emerged the winner of the early 1900s mobility race. The gasoline-powered, internal combustion engine vehicle would dominate the automotive market for decades to come.

- The early 1900s marked a race for dominance between vehicles powered by gasoline, steam(kerosene), and electrical batteries.
- The gasoline-powered, internal-combustion engine vehicle won out thanks to the limited driving range of electric cars, the discovery of abundant oil in Texas, Henry Ford's mass produced, cheaper Model T car, and Charles Kettering's electric starter invention.

ELECTRIC VEHICLES TRY AGAIN

While Ketterling's starter invention, Edison's expensive battery, and abundant Texas oil all helped to halt further progress of electric vehicles in the early 1900s, the electric vehicle did not ever completely disappear. The 1960's themes of Middle East geopolitics, rising energy prices, and OPEC all rekindled interest in the concept of the electric vehicle.

OPEC FLEXES ITS MUSCLES

The 1960s was a time of profound geopolitical change. Countries that had been colonized by developed nations decades earlier were now being granted their independence. As some of these countries attained autonomy, they realized they were in possession of a commodity that their former colonial masters very much craved–oil.

In the early 1960s, Iran, Iraq, Kuwait, Saudi Arabia, and Venezuela moved to establish the Organization of Petroleum Exporting Countries (OPEC). In 1968, OPEC released its *Declaratory Statement of Petroleum Policy* which tersely made clear that member countries had the inalienable right to exercise permanent sovereignty over their natural resources. Membership in OPEC soon doubled as Qatar, Libya, United Arab Emirates, Algeria, and Nigeria joined.

OPEC flexed its muscles when the U.S. government decided to supply weapons to Israel during the 1973 Yom Kippur war. Threats of 5% per month cuts in export volumes sent oil futures prices in the U.S. surging from $3.50 per barrel to over $10 per barrel. Gasoline prices at the pump rose nearly 50% in many American cities.

THE AUTOMAKERS RESPOND

The global energy industry was dominated by the oil-craving Seven Sisters: Gulf Oil, Standard Oil of California, Texaco (now collectively called Chevron), Anglo-Persian (now B.P.), Shell, Jersey Standard, and Standard Oil of New York (now called Exxon Mobil). The Seven Sisters took note of the shifting geopolitical sands and concluded that the oil business had now changed permanently.

Scientists working for these oil companies were tasked with looking beyond oil and refined petroleum as sources of energy for transportation. They were tasked with developing electrochemical storage devices (batteries) that could provide enough energy to move a vehicle. Academic researchers and automotive company engineers soon joined the effort.

In 1966, General Motors introduced its Electrovair II, a Chevrolet Corvair car powered by silver-zinc batteries. Its top speed was around

80 miles per hour and its range was 40-80 miles (depending on how fast it was being driven). The battery pack alone for this prototype cost $160,000, making it unaffordable to consumers and obsolete almost immediately. At the same time, General Motors also introduced its Electrovan powered by a hydrogen fuel cell. Its top speed was 70 miles per hour and its range was 150 miles. Cost concerns ensured this prototype never made it to commercial production.

In 1967, Ford answered the call for electric cars with its Ford Comuta. To save on weight, the body was made of fiberglass. The vehicle had a top speed of 40 miles per hour and a range of 25-40 miles. Only four of these prototype units were ever built.

In 1971, NASA ventured into the electric vehicle resurgence theme with the launch of Apollo 15. Part of the rocket's payload included a battery-powered lunar vehicle called Rover.

In 1975, the American Motor Company produced battery-powered, mail-delivery Jeeps that the United States Postal Service used in a test program.

Despite these various efforts to revive interest in the electric car, the two factors that had dampened its success at the turn of the century remained: a limited top-end speed, and a limited driving range on a fully charged battery.

In 1976, the U.S. Congress passed the *Electric and Hybrid Vehicle Research, Development, and Demonstration Act* which mandated the Energy Department to support research and development for electric and hybrid vehicles. This Act encouraged a new developmental phase. In 1978, General Electric introduced its prototype GE-100, powered by 18, six-volt lead-acid batteries. It had a top speed of 60 miles per hour and a range of 75 miles.

In 1990, General Motors introduced its Impact model. This was General Motors' first attempt at building an electric car right from the design stage. Prior efforts had involved the conversion of existing gasoline-powered car designs to electric power. The Impact had a top speed of 100 miles per hour and a range of about 70 miles. In total, 12 units were built.

In 1992, Ford unveiled its EcoStar van powered by sodium-sulfur batteries. A total of 100 units were built and given to various U.S. companies for use in their corporate fleets. These vans had a top speed of 75 miles per hour and a range of 100 miles. However, each of these units cost around $250,000 to manufacture, making them financially impractical for sale to retail consumers.

As efforts to develop battery-powered vehicles continued, the price of oil reached a peak of near $40 per barrel in 1980 and then started to decline. By 1986, oil prices had fallen to around $11 per barrel; consumers were losing interest in the fuel efficiency argument.

However, sometimes all it takes is a fresh geopolitical spark to restore focus to a theme. In August 1990, Iraq invaded neighboring Kuwait. The U.S. military responded with shock and awe in what was called Desert Storm. The price of oil briefly surged to near $40 per barrel. Once it was realized that the U.S. military had a handle on the situation, oil prices began to decline again. But, the new level of support for oil prices would now be around $20 per barrel. Gasoline prices at the pump took a jump higher with the Desert Storm actions, and never did fall back. The pressure on automotive companies to come up with a battery-powered vehicle intensified.

Honda, Chevrolet, and Toyota all responded to the volatile oil price movements. In 1997, Honda introduced its EV Plus car powered by nickel-metal-hydride batteries. Its top speed was 80 miles per hour and its range was 81 miles. Consumers took notice. A total of 325 units were manufactured and sold.

In 1997, Chevrolet decided to test the small truck market and introduced its Chevy S-10 electric pickup truck. Its top speed was 70 miles per hour with a range of 88 miles. Consumers liked it and 492 units were sold.

In 1997, Toyota entered the picture with a limited release of the Toyota RAV4 EV, powered by nickel-metal-hydride (NiMH) batteries. Its top speed was 85 miles per hour and its range was 95 miles. An impressive 1,484 units were sold. Toyota's encore performance was the Toyota Prius model NWH10, a gasoline-powered hybrid, introduced to Japanese consumers. It was an instant hit. Three years later, in 2000, the Prius model NWH11 was introduced to U.S. consumers.

Thanks to OPEC and geopolitical tensions, the electric vehicle was back, powered by a hybrid gasoline-NiMH combination.

- The 1960's themes of Middle East geopolitics, rising energy prices, and OPEC rekindled interest in the concept of a battery-powered vehicle.
- Over two decades, General Motors, Ford, NASA, American Motors, and General Electric all unveiled prototypes of electric cars. The limiting features of all these designs were the

expensive battery materials, the limited driving range, and the slow top-end speed.

- In the late 1980s, as oil prices receded, consumer's concern over fuel prices subsided.
- In 1990, geopolitical tensions and the U.S. Desert Storm invasion of Iraq set oil on a higher trajectory. Honda, Chevrolet, and Toyota all responded with electric vehicle designs that resonated with consumers. The electric vehicle was back.

HYDROGEN – THE OVSHINSKY ANSWER TO ENERGY STORAGE

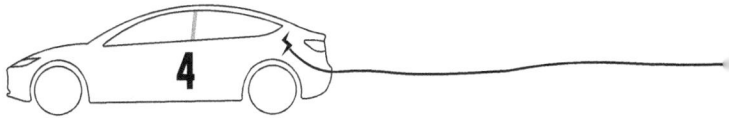

4

Perhaps the reason that the Toyota Prius was a hit with consumers was that it was not purely battery powered and therefore did not need to be charged. The hybrid gasoline-electric design meant that consumers could still gauge the operational efficiency of the Prius in terms of the familiar miles per gallon.

A hybrid vehicle comprises a small gasoline internal combustion engine and a battery pack. Early iterations of hybrid vehicles were designed such that the battery charge level was maintained during driving by the internal combustion engine which powered a small generator and inverter to transform alternating current (AC) power to direct current (DC) power. More recent hybrid design iterations come equipped with a chargeable battery pack and a small gasoline internal combustion engine. The driver charges up the battery before taking the car out on the road. The battery pack will power the vehicle and once nearly depleted of charge, the gasoline engine will be used to get the car to a point where the battery pack can again be charged.

THE NIMH BATTERY

The early iterations of hybrid vehicle were based on the nickel-metal-hydride (NiMH) battery pack. The NiMH battery was patented in 1986 by Stanford Ovshinsky, a self-taught scientist and inventor. Ovshinsky is also responsible for the patenting of flexible, thin-film solar panels, flat screen LCD display screens, and rewritable CD and DVD discs. Throughout his career, he was awarded over 400 patents.

At atmospheric pressure, four grams of hydrogen (H_2) gas occupies 48 liters of volume (the size of a typical car gas tank). This volume of hydrogen gas will release about 472 kJ of energy when combusted. Compare this to gasoline, four grams of which will release 180 kJ of energy when combusted.

Ovshinsky recognized that hydrogen, with its greater energy content, was a desirable fuel for empowering a population fixated on driving cars. But to store enough hydrogen to move a vehicle hundreds of kilometers would mean having a vehicle fitted with high pressure hydrogen storage tanks. His research led him to the idea he termed the *hydrogen ion battery* in which hydrogen energy could be stored in a nickel-hydride alloy within the battery structure, hence the name nickel-metal-hydride battery.

The components of NiMH batteries include an anode of hydrogen-absorbing alloys (the metal hydride), a cathode of nickel hydroxide ($Ni(OH)_2$), and a potassium hydroxide (KOH) electrolyte. The anode is constructed of a complex alloy mixture of the rare earth metal lanthanum, vanadium, titanium, zirconium, nickel, chromium, cobalt, and iron.

During battery charging, the electrolyte dissociates into K^+ and OH^- ions. The OH^- ions react with the ($Ni(OH)_2$) cathode in the following reaction: $OH^- + (Ni(OH)_2) \rightarrow NiOOH + H_2O + $ electrons.

The electrons from this reaction flow toward the anode. The H_2O molecules also flow towards the anode. The reaction that occurs is: Metal + H_2O + electrons $\rightarrow OH^-$ + Metal hydride.

During vehicle operation, the process at the anode reverses itself. The electrons flow towards the cathode, powering the vehicle.

At the cathode, the process is reversed, forming ($Ni(OH)_2$) and OH^- ions again.

Ovshinsky's NiMH design had nearly triple the power density of a standard lead-acid battery. However, the design did have some drawbacks as noted in vehicle testing by automakers:

- Battery performance was prone to underperformance at low temperatures. Not a problem in California, but enough of a drawback to possibly affect vehicle sales in colder regions of North America.
- The battery was prone to self-discharge at warmer temperatures. A car owner leaving his vehicle in his driveway for several days at elevated temperature was likely to find the battery charge had diminished.
- Lower energy density levels meant NiMH batteries would deliver slower acceleration as the driver brought the car up to speed. Faster acceleration would require a bigger,

bulkier, and heavier battery. The extra weight would decrease vehicle performance.

As noted in the previous chapter, the earliest successful attempt at powering a vehicle with a Metal Hydride battery was made by Japanese automaker Toyota. In 2000, Toyota introduced its Prius model NWH11 hybrid vehicle to the U.S. market. It was rated at 22 km/L (52 mpg) in the city. A combination of city and highway driving offered a rating of 40 km/L (95 mpg). Despite the potential battery problems related to temperature, consumers took notice of the vehicle and in calendar year 2000, over 5,000 units were sold.

Honda then countered with a 26 km/L (61 mpg) hybrid vehicle. Ford entered the hybrid race in 2003 with its Prodigy, rated at 30 km/L (70 mpg). But Ford failed to read the consumer correctly; the Prodigy was a diesel-hybrid and consumers could not reconcile their image of dirty diesel fuel with clean hybrid technology. The Prodigy failed to resonate. The marketing lesson learned by Ford and noted by other vehicle makers was one of consistency of brand.

THE CHINA RARE EARTH METAL PROBLEM

Rare Earth metals are the Lanthanide series of elements found on the second row from the bottom in the Periodic Table of the Elements. These elements are called *rare* because they occur in the Earth's crust in very small quantity. Researchers working on battery designs for hybrid vehicles noticed that the use of rare earth metals in the battery anode made for optimum performance. However, China was the dominant supplier of rare earth metals. Fearing that, one day, Chinese geopolitics could lead to rare earth metal shortages, automakers began looking for alternatives to the NiMH design. The fears of the automakers were realized in 2010 when a political spat erupted between Japan and

China over what China felt was an issue of trespass by a Japanese fishing boat in the South China Sea. In retaliation, China threatened to curb exports of rare earth metals to Japan. This geopolitical head-butting dampened enthusiasm for the NiMH battery design for passenger vehicles. Automakers shifted their focus to lithium-ion battery designs. This shift in focus may also have been hastened owing to the fact that a hybrid vehicle has a small gasoline-powered engine. The tailpipe emissions from were at variance with the developing political message of a zero-emissions society.

However, the NiMH battery design has not disappeared entirely. Toyota continues to use this design in its hybrid vehicle models. Geopolitics does however linger in the background in the form of nickel availability. Specifics regarding the NiMH batteries used in vehicles such as Toyota hybrids are difficult to clarify. Based on data in a European publication by battery maker Panasonic, a hybrid battery pack comprised of six modules of 168 individual cells of NiMH construction will weigh around 54 kgs and contain about 8 kgs of nickel material.

- The NiMH battery was patented in 1986 by the American self-taught scientist and inventor - Stanford Ovshinsky.
- Ovshinsky recognized that hydrogen was a desirable fuel for a planet fixated on driving cars. To store enough hydrogen to power a vehicle would require fitting the vehicle with high pressure storage tanks. His research led him to the concept of the hydrogen ion battery where hydrogen energy could be stored in a metal hydride alloy within the battery structure, hence the name *metal hydride battery*.

- The components of NiMH batteries include an anode of hydrogen-absorbing alloys (the metal hydride), a cathode of nickel hydroxide ($Ni(OH)_2$), and a potassium hydroxide (KOH) electrolyte. The anode is constructed of a complex alloy mixture of lanthanum (and other rare earth metals), vanadium, titanium, zirconium, nickel, chromium, cobalt, and iron.
- Ovshinsky's NiMH design had nearly triple the power density of a standard lead-acid battery. However, the design did have some under-performance issues at low temperatures and self-discharge issues at high temperatures. These problems were eventually resolved by using rare earth metals in the anode.
- In 2000, Toyota introduced its Prius hybrid model to the U.S. market. It was rated at 22 km/L (52 miles per gallon) in the city. A combination of city and highway driving offered a rating of 40 km/L (95 miles per gallon).
- The geopolitical barrier to greater market penetration was that China controlled the bulk of the globe's rare earth mineralization.
- In 2010, a political spat erupted between Japan and China and China threatened to curb exports of rare earth metals to battery makers in Japan in retaliation. This was a wake-up call for automakers wanting to use the NiMH battery design and they shifted their focus to lithium-ion battery designs. This shift in focus may have been aided by the fact that a hybrid vehicle with its tailpipe emissions does not align to a zero-emission political vision.
- However, the NiMH battery has not disappeared entirely. Toyota still offers the NiMH battery design in its hybrid models. The NiMH design is not immune to geopolitics thanks to its reliance on nickel metal.

THE BRAINS BEHIND BETTER BATTERIES

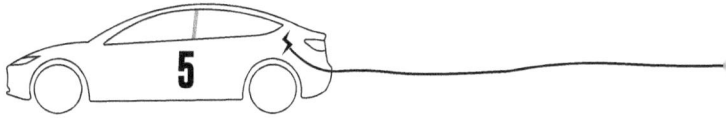

As automakers were experimenting with electric vehicle prototypes in response to rising energy prices, three men were hard at work, focused on improving the chemistry and the performance of batteries that could power vehicles. Their names were: Stanley Whittington, John Goodenough, and Akira Yoshino.

The task of creating a battery-powered car was challenging. Unlike cars from the early 1900s, cars of the 1970s and 1980s were capable of fast acceleration and high cruising speeds. Cars powered by internal-combustion engines had spoiled consumers. The 1970 Chevy Monte Carlo could go from 0 to 60 miles per hour in 8 seconds. Even the Ford Pinto could reach 60 miles per hour from a standstill in 11 seconds. To deliver bursts of power to provide for similar acceleration, a battery would need a fast flow of electrons from anode to cathode. The size of the ion would be the determining factor in its speed of movement from anode to cathode. Adding to this battery development challenge was the fact that even if a suitable battery could be made, there was no infrastructure in place along motorways to provide for battery recharging.

The answer to which ion could be used to create the battery was immediately evident from the Periodic Table of the Elements. The smallest available element that exists in solid form is element #3 – Lithium, with an atomic radius of 0.155 nanometers (0.155 x 10^{-9} meters). However, lithium in a battery would present researchers with a series of scientific challenges to be overcome.

WHITTINGTON AND TITANIUM DISULFIDE

After receiving his PhD from Oxford in the late 1960s, Stanley Whittington moved to California to complete post-graduate studies at Stanford University. His research led him to discover the phenomenon of *intercalation*–the movement of an ion into the crystal lattice of a host material without destroying the crystal lattice structure. He realized that rechargeable batteries (reversible intercalation reactions) depended on: (a) using metals that have crystalline structures, (b) using metals that have empty atomic sites in their crystal lattices, and (c) using metals having good conductivity properties.

Following his studies, he worked for Exxon Mobil on efforts to design a battery that could power a vehicle. In the early 1970s, following on preliminary work done by British scientist Brian Steele for NATO, Whittington designed a prototype battery that used a titanium disulfide (TiS_2) cathode and a lithium-aluminum anode. He discovered that the titanium disulfide material could accept a fast rate of lithium-ion diffusion into its crystal lattice structure. His design proved effective at being rechargeable and delivering enough power to move a vehicle, however, the design was expensive with the titanium disulfide (TiS_2)

material costing $1,000 per kilogram. As well, it was observed in the laboratory that the titanium disulfide material was prone to catching on fire. As a result, Exxon decided not to pursue battery research further. The TiS_2 design was, however, patented. Today's CR2032 coin-shaped battery cells (used in calculators, toys, and vehicle key fobs) are based on Whittington's patent.

GOODENOUGH AND LITHIUM-MANGANESE OXIDE

Whittington's research efforts did not go unnoticed. John Goodenough, a researcher at Oxford University used lithium manganese oxide ($LiMn_2O_4$) as the cathode material instead of titanium disulfide. This battery construct proved amenable to fast charging and high-current discharging. The drawback of this design was one of instability. The battery could not power a high load continuously as the current draw would cause dangerous heat buildup in the battery assembly. Once the battery temperature approached 80°C, the battery could easily catch fire. Even though this battery design was never used in automotive batteries, it was eventually found suitable for use in powering small mobility devices. If you have ever driven a golf cart or a mobility scooter, your ride has been powered by a lithium-manganese oxide battery design.

GOODENOUGH, PADHI, AND LITHIUM IRON PHOSPHATE

In 1986, John Goodenough took up a research post at the University of Texas where he began collaborating with fellow researcher A.K. Padhi. In 1996, they revealed their research on the use of lithium iron phosphate ($LiFePO_4$) as a battery cathode material in combination with graphite anode. This design of battery was shown to have a long life-cycle (over 2,000 charge-discharge cycles), good thermal stability, and was proven able to safely withstand physical abuse. This design of battery is what is used today in many power tool devices. Tesla and Chinese electric vehicle maker BYD have now configured this $LiFePO_4$ cathode design for use in some of their battery designs. The resilience to physical abuse suggests a favorable safety feature in the event of a vehicle crash. The drawbacks of $LiFePO_4$ batteries include a lower energy density and a slower charging rate.

YOSHINO AND LITHIUM-COBALT OXIDE

In the early 1970s, Akira Yoshino joined the Asahi Kasei Corporation in Japan to do chemistry research in support of the Japanese electronics industry that was seeking greater efficiency than what was currently available from nickel-cadmium rechargeable batteries. Goodenough's work caught the attention of Yoshino who used lithium-impregnated polyacetylene material as the anode and lithium-cobalt oxide ($LiCoO_2$) as the cathode material. The drawback of this design proved to be a relatively short battery life-span, low thermal stability and limited load capabilities. Not a problem, however, for small devices. Today this design of battery is found in cameras, laptop computers, and cell phones.

THE NOBEL PRIZE

These three men inspired other scientists in the research community to pursue battery design improvements. Experimentation with manganese and cobalt led some scientists to study lithium-nickel-manganese-cobalt battery designs. The performance parameters of this battery design could be adjusted to suit particular end uses by adjusting the ratio of cobalt, manganese, and nickel in the cathode. Scientists went on to discover that if aluminum was used to replace the manganese, a battery with a high energy capacity could be created. This design would eventually form the basis for the Panasonic NCR 18650B battery which was used in the earliest Tesla electric cars.

Further experimentation with titanium led to the discovery that lithium titanate (Li_2TiO_3) is a suitable material for battery anodes. Tiny crystals of lithium titanate embedded on the anode increase the surface area of the anode and allow for faster electron movement from the anode to the cathode. This battery design is amenable to fast charging and safely delivers a high discharge rate even at low temperatures. The drawback to this battery design was its lower energy capacity which impaired the driving distance of a vehicle using this battery design. The lithium titanate design is now used in laptop computers and in some cell phone models. As well, electric bicycles often use the lithium titanate battery design. The one distinct advantage offered by lithium titanate batteries is the large number of charge-discharge cycles (up to 60,000) which makes for a very long, useful battery life. This feature has not gone unnoticed by the likes of Tesla.

The research of John B. Goodenough, Stanley Whittingham, and Akira Yoshino introduced the world to lithium batteries and changed how the world stores electrochemical energy. In 2019, they were awarded the Nobel Prize in Chemistry for their battery development work.

- Nobel laureates John B. Goodenough, M. Stanley Whittingham, and Akira Yoshino made significant advances into battery designs that used lithium.
- The batteries that power handheld devices and mobility devices are based on their work.

- Their research inspired other scientists to delve deeper into battery chemistry. One of the outcomes of this further research was the lithium-nickel-manganese-cobalt design which Panasonic used to make the batteries for the earliest Tesla electric cars.
- Another outcome was the development of lithium iron phosphate (LiFePO4) battery cathodes which are used by Chinese battery makers today.

THE CLUB OF ROME

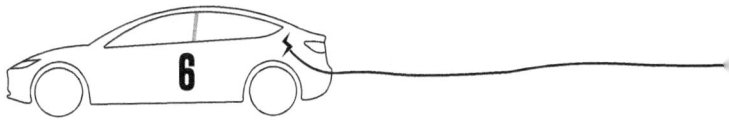

6

Having now examined the history of electric vehicles and the chemistry basics of batteries, let us turn attention to the movement to save the planet and how the electric vehicle theme was swept up into the politics of the environmental movement.

THE PECCEI STORY

The late 1960s were a time of music from names like the Beatles, Elvis, Credence Clearwater Revival, and Otis Redding—to name a few. The three-day rock concert at Woodstock in 1969 defined the counterculture movement. It was the era of peace, love, and rock n' roll.

But behind this party façade, uncertainty was weighing on society. Vietnam war protests were a regular occurrence across the U.S. Memories of the 1962 Cuban Missile crisis lingered. The Cold War between the U.S. and the U.S.S.R. simmered. The average person in

North America had more pressing urgencies than the health of the planet. But in Europe, concerns about the health of the planet were being expressed amongst influential powerbrokers.

One of these influential players was Aurelio Peccei. After graduating from the University of Turin with a degree in economics in 1930, he enrolled at the Sorbonne in Paris for advanced studies. After returning to Italy in the mid-1930s, he was hired by automaker Fiat S.p.A. This was the era of fascism and Benito Mussolini. Peccei was suspected of not being in alignment with Mussolini's fascist ideology. A demotion in the form of a posting to China to promote the Fiat brand of car was designed to discipline him and change his attitude. However, the posting to China proved unexpectedly successful and slowly the political pressure on Peccei eased. Anti-fascist at heart, he had joined an underground anti-fascist resistance movement during the war and for a brief time was imprisoned and tortured. After the war was over, Peccei was freed and he returned to Fiat where he became involved with re-building the war-torn company. In 1949, he took a posting with Fiat to Argentina to help establish an automobile manufacturing plant and restore the Fiat brand to its pre-war status in South America. His career trajectory was on the rise and he was fast becoming known in Italian business circles. He envisioned a future where production processes would be controlled by machines, and a time when data transfer between machines would be commonplace. He even used the expression "artificial intelligence." He was clearly a man ahead of his time.

In 1958, with the blessing of his superiors at Fiat, he founded Italconsult, an engineering and economic consulting group focused on assisting developing countries. Peccei's career trajectory was now unstoppable. In the early 1960s, he became involved with ADELA, an international consortium of bankers aimed at profiting from the spread of industrialization throughout Latin America. Peccei was personally concerned about correcting what he saw as a divide between developed nations and developing nations. ADELA could help him work towards correcting the divide. The ADELA bankers wanted profits. They needed Peccei. Peccei needed them.

Peccei's world view and efforts with ADELA caught the attention of both U.S. Secretary of State, Dean Rusk and vice-chairman of the Russian State Committee on Science and Technology, Jermen Gvishianim. The U.S.S.R. was trying to extend its communist agenda into Latin America and the U.S. was intent on countering these efforts by playing a greater geopolitical role in the region. When Gvishianim decided to invite Peccei to Moscow for private discussions, it became apparent that Peccei was becoming highly influential in political circles.

He next drew the attention of Alexander King, the Director General for Scientific Affairs for the Paris-based Organization for Economic Co-operation and Development (OECD). King and Peccei both agreed that the problems facing mankind were global in nature and required coordinated global action.

THE CLUB OF ROME

The King-Peccei sphere of influence started to grow—attracting engineers, financiers, historians, and visionary thinkers. In 1968, this elite group decided to form the Club of Rome. From the outset, it was agreed that the Club would take a long-term, global perspective and focus on the cluster of intertwined problems they called the *problematique*. Peccei remained involved with the Club of Rome until his death in 1984.

It is fair to say that Peccei shaped the tenets of the Club of Rome, principles which suggest that both developed and developing nations would benefit from more sensible use of energy and other resources as well as a more equitable distribution of global wealth. The Club's tenets further state that people in rich countries will have to change their patterns of consumption and accept lower profits. The paradigm of continuous material growth and the pursuit of endless economic expansion must be challenged.

JURGEN RANDERS

In 1972, King-Peccei follower Jurgen Randers of the Norwegian School of Management attracted the attention of the Club of Rome. Randers philosophy focused on the fact that the Earth was but a sphere in the cosmos. There are no other Earth-like planets that are nearby and easily accessible for exploitation. The mineral and energy resources that exist on planet Earth are finite. Once they are consumed, they are gone forever.

Randers was tasked by the Club of Rome to prepare a report that quantified his finite resource argument. Randers' report stated that in order to feed the 3.6 billion inhabitants of Earth at North American standards of consumption, 0.9 hectares (2.22 acres) of land per inhabitant would be required. To provide housing and infrastructure meeting North American standards, an additional 0.08 hectares (0.198 acres) of land per inhabitant would be required.

Randers' report determined that as of 1972, there were 3.2 billion hectares (7.9 billion acres) of land on the planet suited for agriculture. He estimated that half of this land was under cultivation. Making the remainder of this land suitable for agriculture would require significant capital investment. He went on to argue that if the global population continued to grow, and even if additional land was made agriculture-ready, the world would be facing an arable land crisis by about the year 2000. Randers followed up his report with the book, *The Limits to Growth*, co-authored with academics Dennis and Donella Meadows and William Behrens

The Club of Rome still exists today. Its founding tenets have broadened to now encompass green energy, zero-carbon emissions, and electric vehicles. The mission statement for the Club reads in part: *to apply holistic, interdisciplinary and long-term thinking to ensure broader societal and planetary wellbeing.* This mission and broadened tenets contribute in large measure to the message now being advanced by politicians, the United Nations, the IPCC, and academia.

One section of the Club's website has links to various white paper reports. One title that stands out is: *The Clean Energy Transformation: A New Paradigm for Social Progress.* Consider the following excerpt from this paper: *The impact of the internet and the smartphone on the information sector has cascaded into other sectors as well. There have been rapid improvements in lithium-ion batteries which, in turn, have made electric vehicles (EVs) far more affordable and competitive. Disruptions in the information sector have therefore played a crucial role in the disruption of the global oil industry and a reduction in demand for conventional energy. As costs of EVs plummet, they will make ride-hailing even cheaper than private ownership of internal combustion engine (ICE) vehicles. Autonomous driving technology is improving so much thanks to information technology that it will soon make self-driving cars a reality. This will disrupt private ownership of Internal Combustion Engine cars as Transport-as-a-Service (TaaS) becomes 10 times cheaper.*

Stated succinctly, the Club of Rome does not want you owning a car powered by an internal-combustion engine. They want you hailing a ride in a battery-powered, self-driving vehicle.

In 2019, Oxford University researchers Hannah Ritchie and Max Roser compiled global land usage data. They determined that 4.87 billion hectares of land was being used for agricultural purposes. Multiplying Randers' figure of 0.9 hectares per person by the global population in 2013 (7.2 billion) suggests that 6.48 billion hectares of land should be used for growing food to feed the 2013 global population. This figure shows that agricultural land is being used for other purposes than growing food.

Ritchie and Rosen have calculated the amount of land occupied by housing and infrastructure as of 2013 at 59.5 million hectares. Using Randers' estimate of 0.08 hectares per person being needed for housing and infrastructure shows that the population in 2013 should have used only 46 million hectares, not 59.5 million hectares.

Herein is the problem. In North America, in particular, as the population has expanded and as people's financial situations have improved, housing has spread outward from urban centers. Farmland has been lost in the process. The infrastructure needed to support the outward sprawl of humanity has consumed further amounts of arable land.

The Club of Rome is partly correct. Sensible use of the Earth's resources is required. Living within our means is essential. However, the Club of Rome has erred in its call for disruption of the hydrocarbon energy industry and the discontinuation of automobiles powered by internal-combustion engines. This disruptive stance has contributed to the knee-jerk political reaction favoring electric vehicles.

This electric vehicle knee-jerk reaction constitutes short-term thinking and is not a sustainable solution to a more sensible use of the Earth's resources.

- In 1968, a group of European industrialists and bankers established the *Club of Rome*. The Club vowed to take a long-term, global perspective and focus on the cluster of intertwined societal problems they called the *problematique*.
- In 1972, the Club engaged Norwegian academic Jurgen Randers to write a study exploring these societal issues. Randers penned a worrisome report that argued society was headed for some serious issues by the early 2000s.
- Where the Club of Rome has erred is in its calls for the disruption of the hydrocarbon energy industry and the discontinuation of automobiles powered by internal-combustion engines. This stance has emboldened politicians and policy makers in their resolve to make a shift to electric vehicles.

CHARGED!

THE CARRYING CAPACITY ISSUE

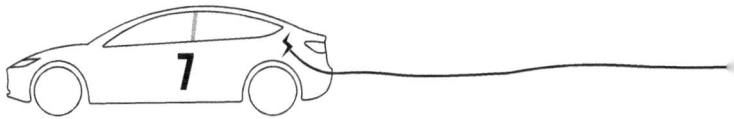

The 1972 ideas of Randers and his colleagues led other academics to consider the idea that the Earth might have only enough resources to carry a defined number of people. This notion is expressed as the *carrying capacity of the planet*.

MALTHUS

The notion of carrying capacity predates the early 1970s. The first person to suggest that the planet might not be able to support an ever-growing population was English economist, cleric, and scholar Thomas Malthus in his 1791 paper titled *An Essay on the Principle of Population as it Affects the Future Improvement of Society.* Malthus was concerned that food production in England would not keep pace with the growing population. He suggested that English society would suffer social turbulence and famine as a result.

In Nature, animals eat what they need to survive. The number of a particular species of animal will be a function of the amount of food available in the ecosystem where the species lives. The carrying capacity of animals in an ecosystem is then the number of animals that can be supported without damaging the ecosystem.

Malthus viewed the world as an ecosystem. He recognized that humans had the propensity to consume more than what was just necessary to survive. He also recognized that humans were a fertile species capable of quickly growing in number. His arguments at the time made sense. Although he did not use the expression *carrying capacity* specifically, what Malthus was hinting at in his writing was that very concept.

What Malthus failed to recognize was the ingenuity of the human species. Post-Malthus, humans have developed technology to enhance food production, to create life-extending medical treatments, to improve hygiene, and to expand global trade. In the context of humans, carrying capacity must recognize more than general survival; carrying capacity must also factor in enhanced food production, life-prolonging medicines, improved hygiene, and global trade that all allow for increased affluence and consumption.

While the world has not ended as Malthus insinuated it might, the population has certainly ballooned. In 1791, Malthus lived in a world with a population approaching 1 billion people. In 2024, the population of the world has reached a plateau of 8 billion people. Although the rate of growth going forward is expected to slow, by 2050 the United Nations expects the population to be over 9 billion. So, what is the carrying capacity of the human population on Earth? A growing number of academics who have entertained this question estimate the capacity at 8 billion people, which is where we are at right now. A small number of academics suggest the carrying capacity to be greater than 8 billion but not exceeding 16 billion people.

IPAT MODEL

Arguments over carrying capacity have been nuanced by economists and environmentalists. In 1971, biologist Paul Ehrlich and environmental scientist John Holdren, unveiled the *IPAT Model* which posits that the impact on the environment is a function of the affluence (consumption) of the planet's population, and technological advances. The IPAT Model indirectly suggests that carrying capacity of Earth is the level at which the impact on the environment does not exceed the ability of the planet to sustain its population. The carrying capacity can rise if people consume just enough to survive. Bigger houses, multiple cars in the driveway, multiple cell phones, multiple laptop computers, and multiple big screen televisions in every household all require raw materials for manufacture. Excessive use of non-renewable raw materials diminishes Earth's carrying capacity.

KAYA IDENTITY

In 1985, Japanese energy economist Yoichi Kaya modified the IPAT Model by factoring in CO_2 emissions. What has come to be known as the *Kaya Identity* suggests that the impact on the environment will only be reduced if energy intensity and/or carbon intensity are reduced. Kaya's work opened the door for the green energy revolution and further added to the theme of electric vehicles.

Taken together, the IPAT Model and the Kaya Identity are a stark reminder that with our current consumption habits, the Earth is at its carrying capacity limit. Continued excessive use of non-renewable raw materials only aggravates the situation.

- As early as 1791, Thomas Malthus began suggesting that there would not be enough food to feed the growing population in England.
- Today there are 8 billion people on Earth and concerns are once more being raised over the ability of the Earth to support us all.
- The IPAT Model originally posited in 1971 is still cited today as the model to follow to determine the Earth's carrying capacity. In 1985, Japanese economist Kaya made the connection between the IPAT Model and CO_2 emissions. This connection has played a role in kickstarting the green revolution.

THE WORLD ECONOMIC FORUM

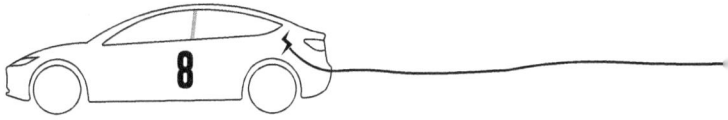

8

The Club of Rome is not alone in its desire for planetary well-being. The World Economic Forum, also started in the early 1970s, today enjoys more visibility and more influence than the Club of Rome.

In the early 1970s, University of Geneva engineering professor Klaus Schwab founded the European Management Forum. This organization focused on the theory of "stakeholder capitalism" as an alternative to conventional business thought. Schwab argued that corporations exist for the benefit of shareholders who have invested in the corporate share structure to earn dividend returns. He argued that under stakeholder capitalism, all individuals impacted by the corporation, including employees, customers, suppliers, and even those impacted by environmental effects should benefit from a corporation's activities.

Taking this idea to a higher level, he started a not-for-profit foundation called the World Economic Forum. He envisioned the foundation to be a global village where corporations and stakeholders could interact, where public-private ventures could take shape, and where strategic insights could be shared with business and political decision makers.

Today, WEF's 28-person Board of Trustees reads like a Who's Who of global decision-making. One-third of the Board are CEOs, one-third are global leaders, and one-third are leaders in civil society. Trustee names include: Al Gore, David Rubenstein (Carlyle Group), Christine Lagaard (European Central Bank), Larry Fink (Blackrock Capital), and Chrystia Freeland (Deputy PM of Canada).

In a 14-minute video on the Forum website (www.weforum.org), Mr. Schwab talks about the history of the Forum. He states that we live in a world where protecting our self-interests has become paramount and in a world that is obsessed by the speed of change. He argues that this mindset runs counter to the idea of treating the planet as a living being.

The Forum (WEF) is now embracing what it calls the *4th Industrial Revolution*. Various white papers posted on the WEF website highlight concerns over net-zero emissions, digital safety, blockchain, the metaverse, food and water security, and climate action. As Mr. Schwab stresses, environmental responsibility is crucial. A holistic approach to regenerate Nature is critical.

Mr. Schwab is correct in his assertion that environmental responsibility is crucial to the future of the planet. However, I think he has been too quick to conclude that electric vehicles are the solution to the problem. The WEF projects that by 2025, the adoption of electric vehicles will have reduced global oil consumption by two million barrels per day. As electric vehicle adoption continues, by 2030 the WEF estimates that 60% of vehicles on the road will be electrically powered and by 2050, society will be at a net-zero emission level. However, the WEF has failed to consider the larger effects on planetary well-being of making such a rapid shift to electric vehicles; I cannot find any evidence on the WEF website where the subject of non-renewable battery raw materials is weighed against the desire to disrupt the method of powering vehicles.

The WEF, in its enthusiasm for electric vehicles, has failed to consider the question—does the planet have sufficient battery raw materials required to bring about widespread adoption of electric vehicles?

- In the early 1970s, Swiss engineering professor Klaus Schwab started a not-for-profit foundation called the World Economic Forum (WEF). His vision was one of a global village where corporations and stakeholders could interact, where public-private ventures could take shape, and where strategic insights could be shared with business and political decision makers.
- The Forum (WEF) is now embracing what it calls the 4th Industrial Revolution with a focus on net-zero emissions, and climate action.
- The stance of the WEF has further encouraged politicians and policy makers in their resolve to make a shift to electric vehicles.

CHARGED!

RANDERS ISSUES ANOTHER WARNING

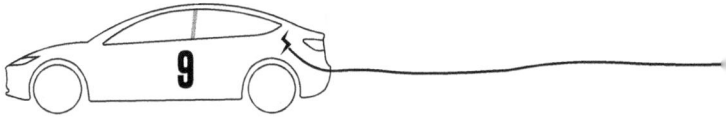

9

Perhaps both the Club of Rome and the WEF should embrace the 2008 and 2013 follow-up findings of Professor Randers. In a 2008 article published in the journal *Futures*, Randers argued that the planet is reaching the point of overshooting its carrying capacity. He warned that if the swelling population causes non-renewable resources to become scarce, the price of these resources will rise, causing a strain on economic growth. The capital cost of growing more food will cause an increase in food prices, which will put a further strain on economic growth. He also argued that these strains on the economy will add to geopolitical tensions around the globe which will act as a further brake on economic growth.

In his 2013 follow-up report to his 1972 paper, Randers described the planet as a closed system with three boundary constraints: arable land availability, non-renewable natural resource availability, and pollution absorption capacity. He argued that bumping up against any one of these constraints will have negative consequences for humanity. For example, consuming arable land for urban sprawl takes away from the available capacity to grow food for the population. Depleting non-renewable resources to create more consumer goods poses limitations to global economic growth. Exceeding the planet's capacity to absorb pollution poses dire consequences for the health and well-being of humanity.

As a prime example of what Randers is referring to, consider that major automakers are creating vehicle battery plants in Canada and the U.S on agricultural land. In his 1972 report to the Club of Rome, he cautioned that he foresaw the day when the world would not have adequate amounts of agricultural land to sustain the population.

General Motors (GM) has announced plans to start making battery packs at several of its vehicle assembly plants. In addition, GM is completing construction of a battery facility at its former CAMI assembly plant which occupies 570 acres (230 hectares) near Ingersoll, Ontario. This geographic part of the province of Ontario is noted for its agricultural output. The use of valuable agricultural land for non-agricultural purposes is precisely what Professor Randers started warning about in 1972.

Ford and South Korean company SK Innovation plan to jointly invest $11.4 billion and create nearly 11,000 new jobs in Stanton, Tennessee, and Glendale, Kentucky to build new electric vehicles and advanced lithium-ion batteries. The Tennessee facility will occupy six square miles of land (1,500 hectares) and the Stanton, Kentucky battery facility will occupy 1,500 acres (600 hectares). More valuable food-producing land consumed. The quoted dollar amount to be invested is somewhat misleading. This figure is actually the amount of tax credit being extended by the U.S. government. In other words, all of the

American taxpayers are ensuring that it will be years before Ford and SK Innovation will have to actually pay wages to the 11,000 employees at these facilities. The money saved on corporate taxes (thanks to the generous tax credits) will cover the wage expenses.

Stellantis has a battery plant in Kokomo, Indiana. The company plans to use tax subsidy money from the U.S. government to build two more facilities. As of mid-2023, the locations of these plants had not been disclosed. In Canada, Stellantis has announced plans to partner with South Korean battery maker LG to build a battery facility on 1,500 acres of prime agricultural land near Windsor, Ontario. The federal and Ontario governments have announced plans to provide $11 billion in subsidies. A total of 2,500 jobs are expected to be created. Stellantis and LG will not have to pay wages at this plant for many years thanks to the generosity of the tax credits paid for by the Canadian taxpayer.

Additionally, Volkswagen has announced plans to build a battery plant in St. Thomas, Ontario with $16.3 billion in government incentives. Battery production will supposedly begin in 2027 and provide thousands of jobs. The Canadian taxpayer is like the gift that keeps on giving. As of early-2024, site preparation is underway on 370 acres of land of agricultural land (the equivalent of 210 football fields).

These various battery plant announcements all have one thing in common. The auto makers make it abundantly clear that 50% of passenger vehicles will be battery powered by 2030.

Randers notes that a key problem in society is the tendency for humanity to engage in short-term thinking. Short-term thinking and desire for short-term consumption is not a new concept. In the early 1800s, short-termism led Scottish philosopher and economist Adam Smith to coin the phrase *the invisible hand*. Randers argues that in order to achieve stable, steady-state economic growth, the global population needs to focus less on the short-term invisible hand and more on regulating the consumption of non-renewable resources, and minimizing the emission of pollution.

Randers' study argues that the size of the global population is slowing and approaching a plateau. Reasons for this include the China one-child policy (1980-2016), women choosing to have fewer children, the general rise in the cost of raising children, and even the shift away from agrarian lifestyles to urban lifestyles. As geopolitical commentator Peter Zeihan reminds his podcast listeners, back in the day when a large portion of the population lived on the farm, children were an asset in that they could help with the farm work. As the family-farm model was exchanged for large corporate farming and as the population gradually shifted to urban dwelling, having several children was no longer financially desirable.

Randers argues that the global population will peak at about eight billion sometime after 2030. As labor availability slows, global economic growth will slow. This, in turn, will ease the urgency for the extraction of non-renewable resources. A levelled-off population will mean fewer pollution emissions.

Randers' bottom line is: there is no pending man-made energy or climate crisis on the horizon. The boundary constraints have been effective in abating population growth. Humanity will survive, albeit at a lower level of per capita wealth and a lower level of per capita consumption.

In the context of electric vehicles, Randers suggests that the knee-jerk reaction towards embracing electric vehicles is overdone.

- In his 2013 follow-up report to his 1972 paper, Randers expresses a unique way of looking at the planet. Randers says the planet is a closed system with three boundary constraints: arable land availability, non-renewable natural resource availability, and pollution absorption capacity. He argues that bumping up against any one of these constraints will have negative consequences for humanity.
- Randers' bottom line is: there is no pending man-made energy or climate crisis on the horizon. The boundary constraints have been effective in abating population growth. Humanity will survive, albeit at a lower level of per capita wealth and a lower level of per capita consumption.
- Randers states that the global population will peak at about eight billion sometime after 2030. As labor availability slows, global economic growth will slow. This, in turn, will ease the urgency for the extraction of non-renewable resources. A levelled-off population will mean fewer pollution emissions.

CHARGED!

THE IPCC
(INTERGOVERNMENTAL PANEL ON CLIMATE CHANGE)

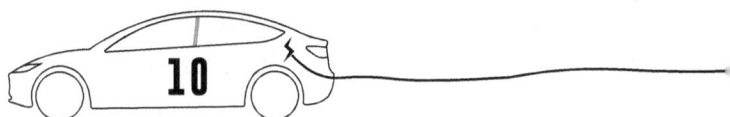

10

In the aftermath of World War II, the U.S. found itself in a position of global political and economic dominance. It was the global power, the global policeman. The U.S. Dollar was the reserve currency thanks to the *Bretton Woods Agreement*. As the war years faded in the distance, the challenge for the U.S. became one of how to maintain this position of power going forward.

UNEP (UNITED NATIONS ENVIRONMENT PROGRAMME)

To meet this challenge, the U.S. government began to rely on the use of scientific assessments to review questions and problems that were arising across various geopolitical fronts. The gathering and parsing of data to help in decision making drew the attention of the United Nations (UN). The UN had become concerned with the environment and had decided that data-based decision making had merit. In 1972, the UN created the United Nations Environment Programme (UNEP) with a mandate to monitor and provide ongoing assessment of the global environment.

As UNEP began undertaking assessments of environmental problems, including climate change, ozone depletion, and loss of biodiversity, it became clear that these assessments and any suggested solutions would have implications for political leaders in various countries. UNEP's aim in its assessments was to achieve scientific consensus among its panel members. This would send a message to political leaders and policy makers that these assessments were worthy of being taken seriously. Any appearance of discord among the scientists working on an assessment had to be avoided lest it undermine public confidence in the experts conducting the assessment and the science being used. Discord also had to be avoided lest it cause political policy makers to delay implementing corrective policy.

UNEP was not acting in a vacuum as it carried out its assessments. It had help from the U.S. military who a decade earlier had begun using satellites to gather data for weather prediction models. This valuable scientific data was seen as aligning well with UNEP's work. The UN called on the intergovernmental World Meteorological Organization (WMO) and the non-governmental International Council of Scientific Unions (ICSU) to collaborate on scientific opportunities that were emerging in the new field of meteorology. The goal was to transform meteorology from a pseudoscience of a forecaster making best guesses on hand-sketched weather maps to a legitimate science involving mathematical models that predicted large-scale, global atmospheric events.

IPCC

The year 1985 marked a major turning point in the meteorological field. Following a conference in Villach, Austria, UNEP, WMO and the ICSU agreed to set up the Advisory Group on Greenhouse Gases. The U.S. government felt that this advisory group, while an important step forward, perhaps lacked the gravitas to fully address climate issues. In 1988, this concern led to the creation of the IPCC (Intergovernmental Panel on Climate Change). With the strong support and influence of the United States, the IPCC announced a mandate to assist the WMO and UNEP in carrying out internationally coordinated scientific assessments to study the potential magnitude, impact, and timing of climate change around the globe. The U.S. government further wanted the IPCC to have an international flavor. This would allow the U.S. government to further expand its foreign policy efforts under the guise of climate change.

The IPCC was given some help in 1992 with the formation of the UN Framework Convention on Climate Change (UNFCCC). Without the UNFCCC, the *Kyoto Protocol* and its agreed-to emission reduction targets would not have been possible. The *Paris Accord* with its goal of holding global warming to within 2°C of pre-industrial levels and pursuing efforts to limit temperature rises to 1.5°C of pre-industrial levels would also not have been possible.

The Paris Accord was a rubicon moment for the IPCC; a point of divergence. There was insufficient hard scientific data to support the 1.5°C and 2°C targets. These numbers amounted to political advice and policy influence. The IPCC was now flexing its political muscle.

The IPCC issued a Special Report in 2018 to add support to the 1.5°C target. The report stated that to limit global warming to 1.5°C, greenhouse gas emissions must peak before 2025 at the latest and then decline 43% by 2030. Crossing the 1.5°C threshold would risk unleashing more severe drought, heatwave, and rainfall events. Limiting warming to 1.5°C implies reaching net-zero CO_2 emissions globally around 2050.

The language in the report immediately throws up a red flag for the reader. The summary section of the report contains language like *estimated, likely, projected,* and *medium confidence.* This was the report that so upset my professor at Heriot Watt University. He completely ignored the nebulous language and focused on the dire prediction for the end of the world as we know it. His reaction to this IPCC Report was a significant factor in my decision to write this book.

The UNFCCC is also instrumental in the annual Climate Change Conference of the Parties (COP) events that bring together politicians, policy makers, and people on the frontline of climate change. If the 2021 COP26 event in Scotland and the 2023 COP27 event in Egypt are any indication, these events will continue to be highly charged media spectacles with the likes of Greta Thunberg and former U.S. Vice-President Al Gore basking in the media limelight. The COP26 event arrived at a consensus that 1.1°C of global warming was due to human activity. There was no data presented to quantify this number. This underscores the political muscle of the IPCC and the UNFCCC. In addition, the conference concluded with an agreement for a phase-down of coal and a phase-out of inefficient fossil fuel subsidies. The COP27 event in Egypt called for developed nations to collectively make $100 billion in annual payments to developing nations that are vulnerable to being harmed by climate change. In 2023, the COP28 event held in the UAE passed policy to replace the $100 billion annual payment sum with a total payment to affected nations of $5.8 trillion by 2030. The COP29 event slated for late 2024 will, ironically, be held in Azerbaijan, one of the birthplaces of the global oil industry. It is expected that food will be a major item on the agenda. The IPCC and UNFCCC are now claiming that one-third of global greenhouse emissions are caused by agriculture. Expect to hear an urgent message to reform the way food is grown, harvested, and distributed. Discussions are also expected to focus on the need to eliminate the $700 billion of annual subsidies to the global agriculture industry. For example, subsidies to U.S. corn growers to provide product to the ethanol industry will come under scrutiny. In fact, expect to hear condemnation for all forms of bio-energy fuels.

The IPCC, UNFCCC, and the COP conference events have now become a source of political manipulation where position statements are not being supported by hard scientific data. Politicians and academics take their marching orders from these COP events and the political push for electric vehicles is a direct consequence.

- In 1972, the UN created the United Nations Environment Programme (UNEP) with a mandate to monitor and provide ongoing assessment of the global environment.
- To lend more political gravitas to this UNEP, in 1988 the IPCC (Intergovernmental Panel on Climate Change) was formed.
- The political punch was strengthened further in 1992 with the formation of the UN Framework Convention on Climate Change (UNFCCC). The UNFCCC is responsible for the Kyoto Accord, the Paris Agreement, and the regular COP events.
- Along the way, the IPCC has morphed into a political influence machine. This is evident in its 2018 Report which does not contain hard scientific data, but yet calls for limits to emissions using language like *estimated, likely,* and *projected future* scenarios.

- The politically oriented IPCC and the COP events are now tools for politicians and policy makers to advance the shift to electric vehicles.

ZEV TAKES CHARGE

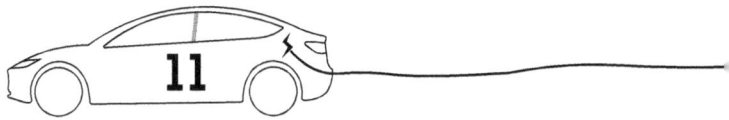

The Club of Rome, The World Economic Forum, and the IPCC are all influential. However, sometimes it takes an entity with legislative gravitas to decisively tip the scales in favor of a new theme. In the case of electric vehicles, that entity was the world's fifth largest economy – the state of California.

CALIFORNIA AND ZEV

In 2002, governor Gray Davis of California implemented *Bill AB 1493* which sought to reduce greenhouse gas emissions from automobiles sold in California, starting with the 2009 model year. The bill laid the framework for tighter emission requirements by way of the *Zero Emission Vehicle Program* (ZEV). This bill even spelled out a direct connection between global warming and exhaust emissions from automobiles.

The National Highway Traffic Safety Administration (NHTSA) initially took issue with California's strategy, as did the Environmental

Protection Agency (EPA). However, in a political about-face, both agencies rescinded their opposition in 2022 which cleared the way for other U.S. states to follow California's model. While ZEV sounds complex, it amounts to a carrot and stick model involving a credit system.

As of 2023, in addition to California, 12 other states have adopted their own variants of the ZEV Program (Colorado, Connecticut, Maine, Maryland, Massachusetts, Minnesota, New Jersey, New York, Oregon, Rhode Island, Vermont and Washington). Collectively, these 13 states are now called the "ZEV states" (zero emission vehicle states). By 2026, Nevada, New Mexico, and Virginia will likely adopt the ZEV program.

Under California's ZEV program, auto manufacturers are classed as either large-volume manufacturers or intermediate-volume manufacturers. Large-volume manufacturers are those who deliver (for sale or lease) more than 20,000 vehicles into California per year. Intermediate manufacturers are those who deliver (for sale or lease) between 4,500 and 20,000 vehicles into California per year. Under the ZEV credit system framework, battery and fuel-cell-powered passenger vehicles delivered to the state for sale or lease are awarded between one and four credits per vehicle, depending on driving range. Hybrid vehicles are awarded 0.4 to 1.3 credits per vehicle, depending on range. Vehicles with internal combustion engines are awarded zero credits. In 2018, California's ZEV program mandated that a total of 4.5% of an automaker's credits had to be from zero-emission vehicles. This threshold will rise to 22% in 2025.

Consider the case of an automaker that each year delivers 100,000 vehicles into the state of California for sale or lease. In 2025, 22% of vehicle credits must be ZEV compliant. At between one and four credits per battery-powered vehicle, this means between 3% and 22% of the company's vehicles delivered into the state must be battery powered. At between 0.4 and 1.3 credits per vehicle, between 17% and 55% of vehicles would have to be hybrid. Should the automaker come up short on its credit count, a financial penalty of $5,000 per credit deficit will be imposed. Manufacturers may bank excess credits earned for future use or trade them with other manufacturers. Manufacturers may purchase credits from other automakers as an alternative to earning them. By 2030, 68% of vehicle credits will have to be ZEV compliant.

CHINA

Beyond the 13 U.S. states that now have the ZEV credit system, the idea has been deemed fashionable in China, with the Chinese adopting the expression *New Energy Vehicle* (NEV) to describe battery-powered, fuel-cell powered, or hybrid units.

The Chinese government will assign each NEV a specific number of credits ranging from one to six, depending on its driving range, battery efficiency, and rated fuel cell system power. The government will then create credit targets for auto companies whose annual production or import volume is 30,000 or more conventional passenger cars per year. Surplus NEV credits can be sold to other companies. Manufacturers unable to meet their credit targets may offset their deficits through purchased credits. China is further considering a credit mandate mechanism for light commercial vehicles as the next phase of its push to see more electric vehicles on its roadways.

EUROPEAN UNION (E.U.)

The European Union has been slow to implement a credit system. The E.U. has decided to instead focus on each automaker's average, fleetwide carbon dioxide emissions. The E.U. has mandated that an automaker's fleetwide CO_2 emissions will be 81 grams per kilometer driven (g/km) by 2025 and 60 grams per kilometer driven (g/km) by 2030. For the period 2025-2029, automakers are encouraged to aim for 15% of their vehicles sold to be battery or fuel-cell powered. By 2030, this sales level will rise to 35%. To incentivize manufacturers to meet the 15% sales target between 2025 and 2029, the E.U. will offer automakers a relaxation in their fleetwide CO_2 emission target by up to 5%. For example, in 2025, if 17% of an automaker's vehicles are battery or fuel-cell powered (2% higher than the target of 15%), its corporate average CO_2 emission target would be eased by a factor of 1.02 (a 2% easing) from 81 g/km to 83 g/km.

CANADA

The Canadian federal government has pledged to meet or exceed its Paris Climate Accord (2015) targets of reducing emissions to 30% below 2005 levels by 2030, and has stated a desire to achieve net-zero emissions by 2050.

Following on these goals, in 2016, the province of Québec passed Bill C-23 to: *reduce the quantity of greenhouse gases and other pollutants emitted into the atmosphere by motor vehicles travelling on Québec roads and so reduce their adverse environmental effects.* Automakers who deliver more than 4,500 vehicles (for sale or lease) into Quebec in a given year will accumulate credits according to parameters, calculation methods, and conditions determined by the government.

In 2019, the province of British Columbia introduced its *Zero-Emission Vehicles Act.* Both British Columbia and Quebec have proposed that by 2035, 100% of vehicles sold or leased must be zero emission.

It seems a foregone conclusion that the remaining provinces in Canada will soon adopt programs that mimic the California credit model. According to Statistics Canada, the share of new registrations of light-duty, zero-emission vehicles in Canada was 2.9% in 2019, 3.5% in 2020, 5.2% in 2021, and 7.2% in 2022. To encourage wider adoption of electric vehicles, the Government of Canada has committed to help build 85,000 publicly-funded charging stations across Canada by 2027.

A BUMPY ROAD AHEAD?

The shift to electric vehicles might encounter a few bumps in the road if the 2021 Chevrolet Bolt recall experience is any indication. In early 2021, General Motors initiated a voluntary recall for 69,000 2017–2019 model year Chevrolet Bolt electric cars citing a risk that manufacturing defects in battery cells might cause the battery pack to catch fire. The recall was based on a reported 19 Bolt vehicles catching fire in the U.S. and Canada. In correspondence sent to Bolt owners, GM asked that while owners waited for a service center appointment to replace the batteries, the owners set the maximum charge limit on their vehicles to 90% and that they not park their cars in a garage. GM's calculations suggested that the total cost of the recall would be over $800 million. But then GM caught a break of sorts. Some owners of Hyundai Kona electric vehicles in South Korea, Canada, Finland, and Austria reported battery fires. It was soon determined that the Kona batteries and the Bolt batteries had been made by South Korean company LG and that the cause of the fires was related to a battery cell tab manufacturing defect. Owners awaiting a service center appointment to have their Bolt batteries replaced panicked and started anxiously clamoring for appointments. GM was facing a crisis. How to promote a shift to electric vehicles and at the same time deal with battery fires? Instead of continuing to replace the batteries on the affected Bolt cars, GM suddenly reneged on the recall and advised Bolt owners that the problem could be solved with software. In correspondence sent to Bolt owners, GM stated: *Dealers will utilize GM-developed diagnostic tools to identify potential battery anomalies and replace battery module assemblies as necessary. The remedy will also include the installation of advanced onboard diagnostic software into these vehicles that, among other things, has the ability to detect potential issues related to changes in battery module performance before problems can develop.*

After deftly sidestepping a potential crisis, General Motors eventually reached an agreement with South Korean LG in which LG picked up the entire cost of battery replacements and software updates. In late 2023, to fend off a possible class action lawsuit, GM offered affected Bolt owners a cash payment of $1,400.

Expect to hear more about zero-emission credit and incentive strategies going forward. Politicians have now convinced themselves that credit mechanisms are the secret sauce that will accelerate the market shift to electric vehicles.

- Government systems of credits, monetary penalties, and fleetwide CO_2 emission target adjustments are designed to incentivize automakers to manufacture more electric vehicles. The leader in devising this carrot and stick approach was the state of California.
- The shift to electric vehicles will not be seamless as the LG battery incident on Chevy Bolt models demonstrated.

THE ELON MUSK FACTOR

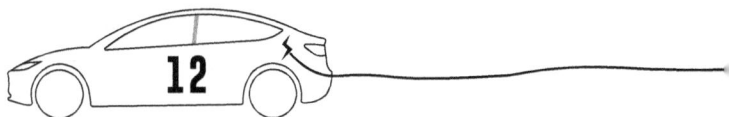

Mention battery powered vehicles in a conversation and the name Elon Musk will immediately pop up. Who is this man and how has he become such a recognized figure?

The Elon Musk story traces its origins to rural Minnesota in 1877 and the birth of his great-grandmother, Almeda Jane Norman. In 1900, she married John Haldeman. Five years later, he developed diabetes, for which there was no treatment at the time. She accompanied him to Minneapolis to see chiropractor Dr. E. W. Lynch, founder of the Chiropractic School and Cure. Chiropractic treatment at the time was a novelty but she was determined to find a cure for her husband's ailment. She took such an interest in what Dr. Lynch was doing to provide relief to her husband that she decided to study under his tutelage and was eventually awarded a certificate in chiropractic studies.

In 1907, the nearby province of Saskatchewan was opening up to homesteaders. John and Almeda made their way to the tiny farming village of Herbert, Saskatchewan where she started a small restaurant and boarding house. She also began offering her chiropractic services to local residents, making her Canada's first female chiropractor.

The couple had two children, Joshua and Almeda Rowena. As a young lad, Joshua suffered a head injury playing hockey at school and his eyesight was affected. His mother took him to Davenport, Iowa to the Palmer School of Chiropractic. The treatments given to him managed to improve his eyesight. This experience made such an impression on him that he later studied at the Palmer School. In 1926, Joshua opened a chiropractic clinic in Regina, Saskatchewan. He married Winnifred Fletcher and the couple went on to have five children. One of these children, Maye Haldeman, was born in 1948.

No doubt inspired by his mother's accomplishments, Joshua had a keen sense of adventure. In 1950, he relocated his wife and five children to Pretoria, South Africa where he established a chiropractic clinic and spent spare time introducing his children to the wildlife in the South African countryside.

In 1970, Maye married Errol Musk, a South African engineer. They had three children: Elon, Kimbal, and Tosca. In 1979, Maye and Errol divorced. Elon decided to live with his father who happened to have a Commodore computer. Elon soon developed an interest in computing and taught himself programming using the BASIC language. At the age of 12, he sold the code for a video game he had created, called Blastar, to the magazine *PC and Office Technology* for $500. With that, Elon was hooked on technology and never looked back. No doubt Elon had inherited his father's sense of adventure. After graduating from high school in 1989, instead of attending university in South

Africa, Elon moved to Canada where he had been accepted into the engineering program at Queen's University.

In 1992, after spending just two years at university, adventure beckoned again and he transferred to the University of Pennsylvania, where, at the age of 24, he received both a Bachelor of Science degree in Physics and a degree in Economics.

In 1995, Elon moved to California to begin studies towards a PhD in Applied Physics and Materials Science at Stanford University. But higher education was not for him. He soon stepped away from the program to pursue his entrepreneurial aspirations in the new, exciting areas of the internet, renewable energy, and outer space. With funding from local investors in Palo Alto, California, he and his brother Kimbal (along with Lebanese-Canadian partner Greg Kouri) started Global Link Information Network, a software company. The software allowed for local Palo Alto businesses to have an online website presence. After a significant cash infusion from an investment fund, the company changed its name to Zip2 and began selling its services to large newspapers who were looking to tap into the power of the new internet technology. Two of the major newspapers that the company signed deals with were the New York Times and the Chicago Tribune. By 1998, the company had sold its services to 160 newspapers across the U.S. In 1999, Compaq Computer paid US$305 million to acquire Zip2. Elon and Kimbal Musk netted US$22 million and US$15 million respectively from the sale. Elon Musk would now be seemingly unstoppable in his quest for new technological startups.

Elon Musk and Greg Kouri wasted little time in starting a new venture. They founded X.com, an online bank that allowed people to send money to one another using email. One year later, the company merged with competitor Confinity, which already had a money transfer service.

The merged entity renamed itself PayPal in 2001. In 2002, PayPal was acquired by eBay for $1.5 billion in shares, of which Musk received $165 million worth.

Musk then turned his attention to his fascination with space flight. In 2002, along with investors, he founded the Space Exploration Technologies Corporation. This company is now commonly recognized as SpaceX. The goal of the company was to develop rockets that could return to their launch pad and land after having journeyed into space.

Not content to sit still, and motivated by legislative efforts in California, Elon Musk next turned his attention to the theme of automotive mobility. In 2003, with partners Martin Eberhard and Marc Tarpenning, Tesla Motors was founded. Over the next four years, over $100 million of financing was sourced from angel investors in California. But the Tesla venture proved to be more difficult than expected. Electric cars were not captivating people's fancy. In early 2008, at the start of the financial crisis, Tesla was in dire shape. In a bid to stave off collapse, Musk pressured Eberhard out along with 25% of the staff, took over as company CEO, and convinced angel investors to come up with a further $40 million to ensure survival.

By 2009, with the financial crisis in the rearview mirror, Tesla had delivered fewer than 150 cars. Finally, the tides began to turn. Consumers began to warm to the idea of electric cars. As Tesla's revenue began to grow, Wall Street took notice. In June 2010, Tesla Motors launched its initial public offering (IPO), which raised $226 million, making it the first American carmaker to go public since the Ford Motor Company in 1956.

Anyone following Elon Musk is aware of his 2022 purchase of social media platform company, Twitter. On the surface, it appears as though this was an ill-fated mistake. However, Musk has demonstrated that sometimes he gets ahead of the rest of the world. The concept of a rocket returning to Earth and using reverse-thrust to land on the landing pad was almost laughable–until Musk made it happen. The concept of a battery powered vehicle was scoffed at–until Musk made it happen. Perhaps he will transform Twitter into something that will amaze all of us. As of mid-2023, he was busy disposing of the Twitter bird logo images and replacing them with images of the letter X which seems to be a point of fascination with him.

- The name Elon Musk is now synonymous with battery powered vehicles.
- Musk has a distinct Canadian connection, dating back to his great grandmother who immigrated to Herbert, Saskatchewan from Minnesota in 1907. She became the first female chiropractor in Canada.
- Musk's grandfather was a chiropractor in Regina, Saskatchewan. In 1950, he set his sights on adventure and moved his wife and five children to Pretoria, South Africa. One of these children was Maye who eventually married a South African engineer by the name of Errol Musk.
- Elon was a precocious child. At the age of 12, he sold the code for a video game he had created to the magazine *PC and Office Technology* for $500.
- In 1992, after spending just two years at Queen's University in Kingston, Ontario, adventure beckoned and he transferred to the University of Pennsylvania, where, at the age of 24, he received both a Bachelor of Science degree in Physics and a degree in Economics.

- Elon, his brother Kimbal, and Lebanese-Canadian partner Greg Kouri, started Global Link Information Network, a web software company, that allowed businesses to have an online presence. In 1999, Compaq Computer paid US$305 million to acquire the company. Elon netted US$22 million from the sale of his shares in the company.
- Elon and Greg Kouri next founded X.com, an online bank that allowed people to send money to one another using email. One year later, the company merged with competitor Confinity, which had a money transfer service called PayPal. In 2002, PayPal was acquired by eBay for $1.5 billion. Musk netted $165 million from the sale.
- Musk next turned his attention to the theme of automotive mobility. In 2003, with partners Martin Eberhard and Marc Tarpenning, Tesla Motors was founded. This adventure would not prove as easy as the previous two. It would take until 2009 before consumers warmed to the concept of an electric car. In June 2010, Tesla Motors launched its initial public offering (IPO), which raised $226 million and made it the first American carmaker to go public since the Ford Motor Company in 1956.

THE CARBON CYCLE

13

Carbon is the fourth most abundant element in our ecosystem behind nitrogen, oxygen, and argon. Carbon atoms form the molecular backbone of ribose sugar molecules which are key to the structure of DNA. Carbon atoms are central to the structure of carboxylic acid which is present in amino acids, the building blocks of proteins. Plant growth relies on carbon dioxide for the photosynthesis process. Carbon in the form of carbon dioxide (CO_2) is a critical part of our atmosphere. Carbon makes life on planet Earth possible.

Planet Earth is a closed system. The planet does not gain or lose carbon. Rather, carbon is in constant movement. This movement, the *Carbon Cycle*, is a naturally occurring process that moves carbon between plants, animals, microbes, minerals in the earth, the oceans, and the atmosphere.

The importance of the Carbon Cycle is ignored by the Club of Rome, the WEF, the IPCC. The message these groups continue to advance is that global warming is bad, emissions are bad, and electric vehicles are good. In fact, all that my Renewable Energy course had to say on the topic of the Carbon Cycle was that the cycle was broken and that the cause was *anthropogenic*–strictly related to the actions of mankind.

INTAKE AND RELEASE

The Carbon Cycle is all about intake and release. Take a deep breath. Hold it for a few seconds and exhale. You have just respired carbon dioxide into the atmosphere. Now multiply that breath by the eight billion breathing people on the planet. The population on Earth is estimated to contribute 29 billion tonnes of carbon dioxide annually just through the actions of breathing in and out. Animals on the planet also breathe and respire carbon dioxide. When plant matter and soil-based organisms decay, the carbon contained in their structures is also partly released to the atmosphere as carbon dioxide. Decaying plant matter that becomes trapped beneath layers of soil over vast spans of time turns into long-chain carbon-based molecules that we otherwise recognize as deposits of oil, natural gas, and coal. When these deposits of oil, gas, or coal are extracted from the ground and combusted in engines or thermal power plants, carbon dioxide is released. When forest fires occur, the burning trees and vegetation cause carbon dioxide to be emitted. When volcanoes erupt, carbon dioxide is emitted. And when people drive vehicles, the combustion of fuel in the internal combustion engines causes carbon dioxide emissions.

Nature seeks a balance. As part of striving for balance, plants absorb this released carbon dioxide from the air during the process of *photosynthesis*. The photosynthesis process uses photon energy from the Sun to chemically combine carbon dioxide with hydrogen and oxygen from moisture in the soil to create sugar-type molecules essential to the hemicellulose and glucose cellular structures of plants, trees, agricultural crops, and grasslands. Humans (and animals) that eat plant material digest the glucose molecules in the plant structure which are broken down in the digestive system and converted into muscular energy.

As part of helping the planet attain a balance, the salty water in the world's oceans also absorbs carbon dioxide from the air. Our oceans contain an estimated 50 times more carbon dioxide than the atmosphere. The CO_2 reacts with the H_2O molecules that comprise the ocean water to form a weak acid called carbonic acid (H_2CO_3). The carbonic acid breaks down to form bicarbonate molecules which are critical to the formation of everything from coral reefs to the hard shells on crustaceans, ranging from tiny sea snails to large lobsters. These reefs and shelled sea creatures make possible the entire food chain for our oceans.

UNBALANCED

The Carbon Cycle is not broken as the academics suggest. Rather, the cycle is unbalanced. This is in part due to the eight billion inhabitants of the planet burning excess amounts of coal, natural gas, and oil to create energy, to power their mobility, and to enhance their economic success. Think back to the earlier chapter where the IPAT model was introduced. The "A" term in the model refers to affluence/consumption. Too many people all striving for mobility and affluence will negatively impact the planet. The next time you are tempted to click your computer mouse and impulse-buy a product online, think about the energy that will be consumed in getting that item to your doorstep. Now think about this purchase in the context of the Carbon Cycle.

The Carbon Cycle has also become further unbalanced from forest fires which emit carbon dioxide. In 2021, it is estimated that 270 million tonnes of carbon dioxide were emitted to the atmosphere from fires. In the spring of 2023, Canada experienced its most serious forest fire outbreak in decades. It is estimated that well over 600 million tonnes of carbon dioxide were emitted to the atmosphere.

Further adding to the imbalance is the removal of vast amounts of forest to produce lumber to build homes for the eight billion inhabitants of the planet. As well, large swathes of rain forest vegetation have also been cut down to make new farmland to grow enough food to feed people. Destruction of forests means fewer trees to absorb CO_2.

Volcanic activity also adds to the imbalance. Just because a volcano is not actively spewing lava does not mean it is defunct. It could still be off-gassing. Measurements taken at Mt. Etna on the island of Sicily show that its off-gassing adds 5.8 million tonnes of carbon dioxide to the atmosphere each year. There are a reported 150 non-active volcanoes worldwide that are off-gassing an estimated 271 million tonnes of carbon dioxide annually. There are an estimated 30 volcanoes worldwide that are periodically active, adding an estimated 117 million tonnes of carbon dioxide to the atmosphere annually. Even seemingly extinct volcanic calderas now filled with water are slowly off-gassing an estimated 94 million tonnes of carbon dioxide annually.

The Earth's crust is comprised of tectonic plates. At mid-ocean locations, these tectonic plates are thermally active and release an estimated 97 million tonnes of carbon dioxide annually. Between volcanic and tectonic action, one can see that Nature is adding nearly 590 million tonnes of carbon dioxide to the atmosphere annually.

Add up the man-made and natural carbon dioxide emissions and contrast them with the absorptions by ocean waters and by plant photosynthesis, and the unbalanced Carbon Cycle comes into sharper focus.

Our atmosphere is a closed system. The carbon dioxide released into the atmosphere accumulates as a blanket in the upper atmosphere which retains heat. As the planet warms, even by a small amount, the solubility of carbon dioxide in ocean waters decreases and the ocean waters release stored carbon dioxide into the atmosphere which further adds to the global temperature rise in a positive feedback loop. The Center for Climate and Energy Solutions states that the global average temperature has increased by about 1°C since the late 1800s with most of the increase coming in the past 25 years. What has not been quantified is how much of this heat comes from people. All eight billion of us radiate heat from our bodies 24-hours a day. Industrial processes and industrial equipment all release waste heat to the atmosphere. In cold-weather climates, energy inefficient houses radiate heat to the atmosphere. If these various sources of heat cause the atmosphere to absorb 1% more heat than it normally otherwise would, that is enough to raise the average temperature of the Earth by an estimated 0.75°C.

Electric vehicles will not stop the bodies of eight billion humans from radiating heat nor will electric vehicles stop industrial processes from releasing waste heat. There are just too many of us on the planet striving for increased affluence. The planet is at or near its carrying capacity. We need to modify our lifestyles. We need to tame our urge for ever-increasing economic materialism and we need to curb our urge for increased mobility.

Driving electric vehicles will not influence volcanic off-gassing. Driving electric vehicles will not influence forest fires. Burning

more coal to generate electricity to charge electric vehicle batteries will not remedy the unbalanced Carbon Cycle.

But even if we do curb our instincts for material affluence, would the temperature of the planet stop trending warmer? Is the warming of the planet exclusive to humans, forest fires, and volcanos? Is it possible that there is something beyond these factors upsetting the Carbon Cycle? Is there something else contributing to the gradual warming of the atmosphere and the ocean waters? Without this mysterious factor, would the ocean waters be capable of absorbing enough carbon dioxide to keep the Carbon Cycle closer to a balanced state?

- Carbon is the fourth most abundant element in our ecosystem behind nitrogen, oxygen, and argon.
- Plants absorb carbon dioxide from the air during photosynthesis, enabling them to grow. Humans and animals eat plant matter which is broken down in the digestive system to create energy. Living beings emit carbon dioxide into the atmosphere. The cycle continues.
- However, the Carbon Cycle has become unbalanced through destruction of excess harvesting of forests and destructive clearing of rain forests to support a growing population. Volcanic eruptions and forest fires have also contributed to the release of carbon into the atmosphere.
- Excess carbon dioxide released into the atmosphere accumulates as an atmospheric blanket that retains heat in our closed system. As the planet warms, even by a small amount, the solubility of carbon dioxide in ocean waters decreases, the

ocean waters release carbon dioxide, a positive feedback loop is formed and temperatures warm further.

- The planet is at or near its carrying capacity. We need to modify our lifestyles. We need to tame our drive for ever increasing economic materialism. We need to curb our urge for increased mobility.
- There may be another factor beyond humans, forest fires, and volcanos that is influencing climate change and upsetting the Carbon Cycle.

CHARGED!

CLIMATE CHANGE IS CYCLICAL

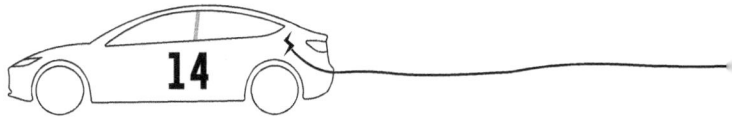

In my introductory geology course during my first year of engineering studies in 1982, we learned about the structural models that explain the Earth's geologic features. The textbook that I used in this course contained only one short paragraph mentioning climate change and how it was related to cyclical variations in the orbit of Earth around the Sun. At no time did the professor even mention climate change; it was not something that people were focused on in 1982.

MILANKOVITCH

In 1976, authors Hays, Imbrie and Shackleton published an article in the journal *Science* entitled *Variations in the Earth's Orbit: Pacemaker of the Ice Ages*. The authors were granted access to drill-core material taken from two locations in the southern Indian Ocean. This core material had been obtained as part of a multi-nation 1970s study called *Climate: Long Range Investigation, Mapping, and Prediction* (CLIMAP).

In their analysis of the drill core material, Hays, Imbrie and Shackleton focused on the *Milankovitch Cycles*. These cycles were proposed by Serbian astrophysicist Milutin Milankovitch. Born in 1879, Milankovitch graduated from Vienna Technical University in 1904 with a doctoral degree. He spent his entire career researching and teaching at Belgrade University in Yugoslavia (modern day Serbia). Building on the fact that Earth is tilted on its orbital axis, Milankovitch suggested that the angle of tilt varied from 22.1 to 24.5 degrees over a cycle of about 40,000 years. He further suggested that the Earth orbits the Sun in an eccentric pattern that is more elliptical than circular. He suggested that the orbital pattern varied from high eccentricity (nearly elliptical) to low eccentricity (nearly circular) and back again over a long cycle of around 100,000 years. Lastly, he noted Earth rotates (spins) on its axis with a slight wobble. The cycle of this wobble (called the *Precession of the Equinox*) measures 25,772 years in duration.

What Hays, Imbrie and Shackleton concluded from examining the drill core data was that climate change is cyclical. Their data analysis pointed to three long cycles that overlap with one another in a complex fashion. Their data analysis showed these long cycles were very closely aligned to the cycles proposed by Milankovitch.

In other words, the climate of Earth is always changing as a result of these long cycles.

But the rate of change is variable thanks to these long, overlapping cycles. Over the past 10,000 years, the climate has been stable, with average global temperature ranging between an estimated +/- 0.4 °C relative to the current 14°C average global temperature. Ice core data from Antarctica shows that this recent 10,000-year stable, warm period is an exception. Up until about 12,000 years ago, the planet was in a cold glacial period with temperatures cold enough to allow polar ice to extend across much of northern Europe and North America. The ending of this period is what geologists refer to as the transition from the Pleistocene epoch to the Holocene epoch. Ice core data further reveals that these glacial periods are historically the norm. About 80% of time in the last 800,000 years the planet has experienced glacial periods with temperatures 4 to 8°C colder than the current average of 14°C.

INTERGLACIAL PERIODS

Only every 100,000 years or so do warmer periods of about 10,000 to 15,000 years occur. The planet is currently in one of these inter-glacial periods. This period could end anytime, or it could continue for another few thousand years. We have no way of knowing.

An inter-glacial period unfolds as follows:

1.   Warming starts because of changes in how Earth rotates on its axis and orbits around the Sun. Every 100,000 years or so the planet is just in the right position for the northern part of the globe to get considerably more sunlight (northern Europe, Russia, Greenland, Arctic, Canada, northern United States).
2.   The warming melts large parts of the ice sheets that cover the terrain. The melting ice releases warm fresh water which changes the ocean currents and the ocean temperature.
3.   The warmed ocean water cannot retain as much carbon dioxide in solution and the oceans become a net emitter of carbon dioxide. The large amounts of released carbon dioxide result in more atmospheric warming. The planet enters a warmer period which lasts 10,000 to 15,000 years.
4.   As the Earth's orbital pattern continues to change during the 10,000 to 15,000 or so warmer years, the Northern

Hemisphere starts to receive less sunlight and becomes gradually colder. Snow and ice are more frequent. Snow and ice reflect more sunlight than bare earth and the net result is a colder climate.

5. A glacial period begins when large amounts of added carbon dioxide from this earlier warmer release get re-absorbed into the cooler ocean waters. The glacial period ends when the Earth's orbit is again just in the right position to let the Sun melt the majority of the vast sheets of ice that have formed in the north.

The above steps than repeat.

To further appreciate climate change, stop for a moment and consider that there are oil and gas deposits in the North Sea off the coast of Norway. There are massive untapped oil and gas deposits in northern British Columbia that extend into the adjacent North West Territories. For these deposits of hydrocarbons to have developed, millions of years ago these areas must have been warm and wet with plenty of decaying vegetation. Scientific drill core data indicates that about 49 million years ago, the planet was some 12°C warmer than today. Mankind was not around millions of years ago to cause these warm conditions. This is further evidence that variations in climate follow the Milankovitch cycles.

The population of Earth in the early 1900s when Milankovitch was studying these cycles was around 1.6 billion people. Today, the population of Earth is at the eight billion mark. Although there is the possibility that climate change is a combination of Milankovitch cycles and anthropogenic (man-made) factors, the extent to which cyclicality and anthropogenic factors each are contributing to climate change cannot be quantified with accuracy. Scientists do not have enough long-term data to make these distinctions. The first reasonably accurate thermometer was not invented until 1715. It was not until around 1880 that scientists started to accumulate and save temperature data on a regular basis. The amount of available temperature data is less than 150 years, worth. To put this brief time span into context, the last glacial period ended some 12,000 years ago. Our planet is over four billion years old. So, when the IPCC or a COP Conference offers temperature numbers to support policies and resolutions, there really is not enough hard data to support their ideas. Yet, the academics and the politicians continue to believe what these groups have to say.

Mankind will not be able to alter the course of the Milankovitch cycles, despite what the Club of Rome, the World Economic Forum, the IPCC, academics, politicians, and media say.

Ignoring the long cycles of Nature and thinking that mankind is solely responsible for climate change is short-sighted and naive.

The knee-jerk reaction to promote electric vehicles will not influence the larger Milankovitch cycles.

Having now looked at the Carbon Cycle, Milankovitch cycles, a brief history of electric vehicles, and the influencial groups that are promoting electric vehicles, let's delve deeper into the materials used to make lithium-ion batteries. This examination will further challenge the thesis that electric cars are the answer to what ails the planet.

- Serbian astrophysicist Milutin Milankovitch (1879-1958) suggested that the axial tilt of planet Earth varies from 22.1

to 24.5 degrees over a cycle of about 40,000 years. The Earth orbits the Sun in an eccentric pattern that is more elliptical than circular. Variations in this eccentricity unfold over a long cycle of around 100,000 years. The Earth rotates (spins) on its axis with a slight wobble; the cycle of this wobble measures 25,772 years in duration.

- A 1976 study of ice core material by researchers Hays, Imbrie and Shackleton confirmed the validity of the Milankovitch cycles.
- The climate of Earth is always changing. But the rate of change is variable thanks to these long overlapping cycles.
- About 80% of time in the last 800,000 years, the planet experienced glacial periods with temperatures 4 to 8°C colder than the current average of 14°C. Only every 100,000 years or so do warmer periods of about 10,000 to 15,000 years occur.
- The extent to which both these cycles and anthropogenic factors contribute to climate change cannot be quantified with accuracy. Scientists do not have enough long-term data to make these distinctions.
- Mankind will not be able to alter the course of the Milankovitch cycles, despite what the Club of Rome, World Economic Forum, the IPCC, academics, politicians, and media say. Ignoring the long cycles of Nature and thinking that mankind is solely responsible for climate change is short-sighted and naïve.

THE BASIC DESIGN OF A LITHIUM-ION BATTERY

15

The basic science behind a lithium-ion battery is similar to the basic science behind the lead-acid battery. A lithium-ion battery is comprised of two adjacent cells, separated by a membrane and some electrolyte material.

In one cell is the electrode (anode), constructed of graphite impregnated with lithium. The chemical nomenclature for this anode material is *lithiated graphite* (LiC_6). In the other cell is the electrode (cathode) made of a fused collection of metals in oxide form (nickel oxide, cobalt oxide, manganese oxide, lithium oxide). Each cell further contains a charge collector plate made of either a thin layer of aluminum or copper material.

The electrolyte material is a liquid or paste comprised of lithium hexafluorite (LiF_6). However, some batteries use lithium tetrafluoroborate ($LiBF_4$) which has greater thermal stability.

The membrane material separating the two halves of the battery cell is generally a lithium-phospho-fluorite (LiPF$_6$) compound embedded in a propylene carbonate or diethyl carbonate material.

The lithium-ion battery is based on the science of intercalation–the ability of a material to allow ions move in and out of its crystal lattice structure. Because the lithium ion is so small, it can move into the atomic structural vacancies in the lithiated graphite anode and metal oxide cathode material.

In the cell containing the anode, the chemical reaction is:

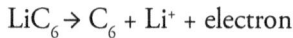

$$LiC_6 \rightarrow C_6 + Li^+ + electron$$

In the cell containing the cathode, the chemical reaction is:

$$metal\ oxide + Li^+ + electron \rightarrow Li \cdot metal\ oxide$$

When the battery is charging, Li$^+$ ions *and* electrons move from the cathode to the anode, where they both collect. When the battery discharges, to power an electric motor that drives a vehicle, the Li$^+$ ions *and* the electrons generated at the anode flow towards the cathode where they both collect. This movement of electrons is what is known as *electricity*.

The separator barrier between the adjacent cells allows the Li$^+$ ions to move from cathode to anode (and back again) while preventing electrons from also coming along. The electrons are forced to move via the external connective wiring joining the anode and cathode.

Lithium is unique in that it has a high degree of *electropositivity*. That is, a lithium atom gives up its electrons very easily. When a driver wishes to make an electric vehicle move faster, the ability of lithium to donate its electrons quickly is what is responsible for the burst of power that turns the vehicle wheels faster.

A typical cell arrangement in a lithium-ion battery will generate a voltage of around 3.7 volts (+/- 0.5) depending on the exact composition of anode and cathode material. The capacity of a typical cell is around 3 amp·hrs. You might recall from high school physics class the experiments to demonstrate the effect of connecting batteries in series and parallel. For example, two 12 volt batteries (100 amp·hr capacity) connected in series will provide 24 volts with 100 amp hrs of capacity. Those same two batteries connected in parallel will provide 12 volts and 200·amp hrs of capacity. Electric vehicle battery makers have resorted to a combination of series and parallel connections by assembling the individual cells that have been connected in series into module groupings that are connected in parallel.

To illustrate the effect of using a combination of series and parallel connections, consider that a 12-volt lead-acid battery with six cells of 2 volts per cell connected in series (6 x 2 volts = 12 volts) will have a capacity of about 20 amp·hrs. The energy in the battery will be 0.24 kW·hrs. The battery will have enough energy to start the vehicle and supply spark to the cylinders, but will require recharging from the alternator as the vehicle drives. In the absence of recharging, a load of 2 amps will cause the battery to deplete its charge in 10 hours.

Consider a battery pack in an electric vehicle capable of producing 400 volts. The capacity of the battery pack is 240 amp·hrs. The energy of the battery pack is 96 kW·hrs (400 x 240/1000). At highway speeds, suppose the electric motors driving the vehicle are drawing a load of 60 amps. The battery will have enough energy to move the vehicle for four hours. The goal of the automaker is to have the battery provide power to the vehicle drivetrain for as long a time as possible. Driving range is a key selling feature for electric vehicles. Battery makers are thus looking to optimize the cathode and anode chemistry and to optimize the battery weight.

Now, let's take a deeper look at the battery materials and the design iterations that battery makers have gone through.

- A lithium-ion battery is comprised of two adjacent cells, separated by a membrane and some electrolyte material.
- In one cell, is the electrode (anode) constructed of graphite impregnated with lithium (LiC_6). In the other cell is the electrode (cathode) made of a fused collection of nickel oxide, cobalt oxide, manganese oxide, and lithium oxide.
- In the cell containing the anode, the chemical reaction is: $LiC_6 \rightarrow C_6 + Li^+ + electron$
- In the cell containing the cathode, the chemical reaction is: metal oxide + Li^+ + electron $\rightarrow$ Li·metal oxide
- When the battery discharges in order to power an electric motor that drives a vehicle, the Li^+ ions *and* the electrons generated at the anode flow towards the cathode. This movement of electrons is what is known as electricity.

BATTERY MATERIALS AND DESIGN ITERATIONS

Academic and political proponents of electric vehicles have failed to thoroughly consider the availability, future supply, and environmental impact of the metals, materials, and chemicals used in making lithium-ion batteries. Use of these materials will have a critical bearing on the environmental health of the planet and on the planet's ability to provide raw material for future generations. Proponents of electric vehicles have further swept the technical descriptions of batteries off to the side. The average consumer is being offered a picture of all electric vehicle batteries being equal.

NASTY CHEMICALS

Let's consider the chemicals used in the process of lithium battery manufacturing. The electrolyte material in a lithium battery is created with a mixture of ethylene carbamate (urethane) and lithium hexafluorophosphate ($LiPF_6$). To create $LiPF_6$, the starting point is a reaction of lithium carbonate (Li_2CO_3), lithium fluorite (LiF), and hydrofluoric acid (HF). The resulting slurry is then combined with ammonia (NH_3) to bring the pH to a neutral level of 7.5. The slurry is then dried. Phosphorous pentachloride (PCl_5) is then blended with the dried slurry at a temperature of 78°C. Hydrofluoric acid (HF) is added to bring about a reaction whereby the Cl^- ions are exchanged for F^+ ions. The end result is $LiPF_6$.

This complex set of reactions begs the question: what is the toxicity of these various chemicals? Hydrofluoric acid is nasty material that cannot be neutralized and disposed of down the sewer. Trying to neutralize any waste HF by blending with a base material (pH greater than 7) will generate toxic metal salt compounds. The Material Safety Data Sheet (MSDS) for HF cautions that it must be isolated in special sealed containers and handled by a waste management company. The MSDS sheet for ethylene carbamate states that it too is a nasty substance that requires careful handling and disposal. Lithium hexafluorophosphate material is also toxic and the MSDS sheet spells out careful disposal measures. Phosphorous pentachloride is so toxic it is potentially fatal if inhaled.

Making the separator membrane that isolates the cathode part of the battery from the anode part typically starts with a film of polyethylene material. It is coated with dibutyl phthalate (DBP) which has been dissolved in acetone (C_3H_6O). The coated layer is then heated to 130°C to evaporate off the acetone. The MSDS sheet for DBP states that it is very toxic to aquatic wildlife and can impair the fertility of people who handle it. It is to be treated as a hazardous waste.

The unsettling part of these chemicals is that they are very often manufactured in nations such as Slovakia, China, and India. The question that lingers in my mind is one of safety. Are the employees involved in processing these materials doing so in a manner that is safe to their health? Is the residual waste material being handled in an environmentally-safe manner? Are residual process vapors being vented to the outside atmosphere at the manufacturing plants?

If you buy an electric vehicle, are you contributing to health issues of people in foreign countries?

BATTERY CELLS – POUCH DESIGN

Let's next take a look at the design evolutions of lithium-ion batteries. In 1995 the pouch battery design was introduced for lithium batteries used in laptop computers. The anode, cathode, separator, and electrolyte materials are encased in an aluminum-coated, plastic film pouch. Take apart your laptop computer and you will see what a pouch design looks like. General Motors has collaborated with South Korean technology company LG to make pouch-style batteries for its Chevy Bolt, Cadillac Lyriq, Buick Electra, and Hummer EV models. This battery design has been branded the *GM Ultium* battery. A single pouch measures about 580 mm x 113 mm x 10 mm thick. The number of pouches stacked together will determine the overall battery capacity. For example, the Hummer EV has a battery assembly with 212 kW·hrs of capacity and a total battery assembly weight of 1,326 kgs. The major drawback to the pouch design is that after around 500 charge-discharge cycles, the pouches will usually exhibit swelling of up to 10% in volume. This feature raises safety concerns.

LG's involvement with pouch-style batteries predates the General Motors decision to enter the electric vehicle market. In 2016, LG unveiled its LGX E63 battery design. A pouch measured 125 mm x 325 mm x 11.5 mm thick and weighed 965 grams. The 2016 Renault 20E car used around 170 stacked pouches in its 41 kW·hr battery pack.

This was followed with the similar-sized LG E66A battery which was used in the first models of Chevy Bolt cars.

BATTERY CELLS - CYLINDRICAL DESIGN

In some lithium-ion batteries, the individual cells are of a cylindrical design. This type of design is not a new concept. For example, the AA batteries that you use to power a small flashlight are a cylindrical design. Make the cylinders bigger and bundle together hundreds of them and you have the basis for a lithium-ion battery pack that will power a vehicle.

To make an individual cylindrical cell, thin foil sheets of anode, separator, and cathode material are sandwiched, rolled up, and packed into a cylindrical shell made of thin-gauge steel. A small tab connects the anode and cathode material to the wall of the cylindrical shell.

THE 18650 CYLINDRICAL DESIGN

One early design of cylindrical cell was the 18650-type. Each cell measures 18 mm in diameter and 65 mm in length (hence the 18650 nomenclature) and weighs approximately 47 grams. Each cell produces a voltage of 3.7 volts, and stores up to 3.5 amp-hrs. The innovator responsible for this design is Japanese electronics manufacturer Panasonic. A significant number of electric vehicle makers are using the 18650 design. The main drawback is the relatively small 3.5 amp-hr cell capacity. To provide for a decent driving range, a large number of cells must be assembled into the battery pack design. For example, a Tesla vehicle fitted with a type 18650 battery has a reported 7,920 cells assembled into five modules of 1,584 cells per module. The weight of a battery pack with 7,920 cells will be at least 375 kgs. Tesla has collaborated with Panasonic at the Tesla Nevada battery plant to produce 18650-type cells.

2170 CYLINDRICAL DESIGN

Another early design of cylindrical cell was the 2170-cell. Each cell measures 21 mm in diameter and by 70 mm in length and weighs approximately 68 grams. Each cell produces a voltage of 3.7 volts, and stores up to 4.8 amp-hrs. This storage is a significant improvement over the 18650 design. The innovator responsible for this design is family-run, South Korean electronics firm LG (formerly known as Lucky Goldstar). The increased dimensions of this design mean electrons have to travel further along the anode material to reach the tab bridging to the cathode material in order to complete their electrochemical reactions. This results in increased amounts of heat being generated in the battery and therefore vehicle designers have to apply extra engineering efforts to ensure the battery pack in the vehicle remains sufficiently cool during vehicle operation. These design efforts add to the overall selling price of the vehicle.

Some earlier versions of Tesla cars used the 2170-type battery. Some models used a 2170 design with 2,976 cells while other models used a 2170 design with 4,416 cells. Given the higher storage capacity of this design, a battery with 4,416 cells weighs less than a battery pack made of the 18650 design. Considering that each cell has a pair of electrical connections, a total of 4,416 cells requires 8,832 connections. Compare this to an 18650-type design with 7,920 cells which will have 15,840 connections. If each connection represents a potential failure point as the vehicle battery ages, the probability of a battery failure diminishes in a design with fewer connections.

PRISMATIC DESIGN

Prismatic cells are made in a manner similar to cylindrical cells, except the cylindrical cell is partly flattened to resemble a prismatic (hexagonal) shape. More prismatic cells than cylindrical cells can be packed into a unit of battery volume. There is no standard sizing nomenclature for prismatic cells. They can be made as large as 5 kgs in weight to accommodate the design parameters of an electric vehicle manufacturer. A single prismatic cell can be made to contain as much energy as perhaps 100 cylindrical cells. The innovator responsible for this design is Chinese firm Contemporary Amperex Technologies Company Limited (CATL). Founded in 2011, CATL now has around 40% of the global electric vehicle battery market and has 13 manufacturing hubs around the world, including factories in Hungary and Germany.

Many Chinese electric vehicle makers have opted for the prismatic design. Electric vehicle makers in other countries have not been so fast to embrace this format. The reason is CATL has based its prismatic cell design on cathodes made of thin foils of lithium iron phosphate (LiFePO$_4$). This design decision means that the battery will deliver slower energy discharge, which means slower vehicle acceleration. It will be interesting to see if this battery format ever arrives in North America. Will North American consumers forgo their need for rapid acceleration?

4680 CYLINDRICAL DESIGN

To improve upon the 18650 and 2170 battery designs, Tesla engaged the help of a Canadian researcher, Dr. Jeff Dahn of Dalhousie University. What Dahn and his team developed was the idea of laser-etching tiny holes into thin foils of either nickel-cobalt-aluminum oxide material or nickel-cobalt-manganese oxide cathode material. As with prior battery designs, Dahn and his team retained the anode design comprising a thin foil of graphite power coated onto a copper backing. Dahn's team showed that the tiny etched holes in the cathode reduced the length of the pathway the electrons have to follow. They proved that the etched-hole design delivered a reduction in the amount of heat generated during battery discharge. Dahn's idea is now referred to as the "tabless-design" because electrons do not have to travel through the battery tabs in order to discharge.

Tesla has taken Dahn's design work and incorporated it into cells measuring 46 mm in diameter by 80 mm in length (hence the 4680 nomenclature). Each cell weighs about 355 grams, produces a voltage of 3.7 volts, and stores up to 26 amp-hrs. This increase in cell size together with the laser-etched hole design creates a cell with 14% greater energy density. Greater energy density translates directly to a greater driving range for the vehicle.

To put these design advancements into perspective, consider a 2170-type battery design with 4,416 cells weighing 300 kgs. Consider also an 18650-type battery design with 7,920 cells weighing 375 kgs. At an individual cell weight of 355 grams, a battery pack based on 4680-type cells will have an estimated 960 cells and will weigh about 340 kgs. The reduced number of cells in the 4680-type design translates into reduced chances of battery pack failure. The major benefits to the consumer, however, are increased acceleration and a reported 16% greater driving range versus the 18650-type design. The Tesla battery factory in Texas is focused on the 4680-design and a press release in late 2023 announced the completion of the 20-million[th] 4680-type battery. However, it has also been reported (based on information provided by insiders) that Tesla is having technical challenges with the 4680-type battery. At issue is the fact that this design utilizes dry-coated cathode materials instead of the more traditional wet design which requires drying prior to cell assembly.

CATHODE MATERIALS

No matter the design of the battery, it will need a cathode. Precise details on battery cathode designs are hard to find and for good reason —battery makers are seeking to safeguard their design details. What is not a guarded secret is that the materials used for the battery cathodes are all non-renewable resources. Once they are mined from the Earth and consumed, they are gone forever.

The question is: will the electric vehicle industry be able to modify its cathode designs to supply the number of electric cars envisioned by political policy makers, while at the same time safeguarding the future supply of these non-renewable materials?

NCA AND NCM CATHODE DESIGNS

Lithium-ion batteries used in vehicles sold in North America and Europe currently employ one of two different cathode designs: lithium-nickel-cobalt-aluminum (NCA) or lithium-nickel-cobalt-manganese (NCM). Batteries with NCA or NCM cathodes are known for having a high energy density, very good driving ranges, and customer-pleasing acceleration. NCA and NCM designs are reported to be good for up to 2,000 charge-discharge cycles.

622 NCM Design
South Korean technology company (LG) was an early mover in the lithium battery sector. One of the earliest NCM cathode designs it trialed was the *622 architecture*. The ratio of the cathode elements

was 6:2:2. The stoichiometric formula of the cathode alloy was $(Li \cdot Ni_{0.6} \cdot Co_{0.20} \cdot Mn_{0.20})O_2$.

800 NCA Design
LG also developed an NCA cathode design using nickel, cobalt, and aluminum. In this design, the ratio of the cathode elements was 8:1.5:0.5. The stoichiometric formula of the cathode alloy was $(Li \cdot Ni_{0.8} \cdot Co_{0.15} \cdot Al_{0.05})O_2$. Low cycle life concerns prompted LG to abandon this architecture in favor of NCM cathode designs. Meanwhile, Japanese technology company Panasonic also developed a battery based on the NCA design. The first Tesla Model S cars released in 2012 used this Panasonic NCA architecture in the batteries.

South Korean LG was not without domestic competition. South Korean technology firm, SK Group, decided to enter the battery sector with an NCM architecture of 9:0.5:0.5. The stoichiometric formula of the cathode alloy was $(Li \cdot Ni_{0.9} \cdot Co_{0.05} \cdot Mn_{0.05})O_2$. One of the challenges of raising the nickel content in the cathode is a reduction in battery life. SK Group was able to remedy this situation by reducing the amount of cobalt and manganese in the alloy blend. In 2019, Tesla moved away from the NCA battery design to the NCM architecture on vehicles it produced at its assembly plant in China. However, Tesla did retain LG as the battery maker.

811 NCM Design
The cathode material in a 4680 battery cell is reported to be 81% nickel based. The key word is *reported*, because Tesla has been tight-lipped about precise battery details. It is thought that the 4680 cathode is based on what is called 811 chemistry, a term derived from the stoichiometric formula $(Li \cdot Ni_{0.8} \cdot Co_{0.10} \cdot Mn_{0.10})O_2$. This formula implies that for every mol of nickel used, the amount of lithium used is 1.25 mols.

MATERIAL USED IN THE 4680 DESIGN

An individual 4680-type cell weighs 355 grams. Precise details on the structure of individual cells are guarded, though we can do some mathematical calculations to learn more.

The thickness of the cell cylindrical can structure is about 0.9 mm. A cell can (including the top and bottom) in a 4680-type battery will weigh just shy of nine grams assuming the flat-rolled steel has a density of 7.8 grams per cubic centimeter. The anode material will weigh around three grams. This means the cathode material in the cell can will weigh around 343 grams.

The stoichiometric formula of $(Li \cdot Ni_{0.8} \cdot Co_{0.10} \cdot Mn_{0.10})O_2$ means that 1 mol of Li-Ni-Co-Mn material is matched by 1 mol of O_2 which weighs 32 grams. Therefore, the Li-Co-Mn material must account for the remaining 311 grams of the cathode cell weight. The gram molecular weights of these components are: lithium 6.94 grams per mol, nickel 58.69 grams per mol, cobalt 58.93 grams per mol, and manganese 54.93 grams per mol. A total of 311 grams of material with a stoichiometric formula of $Li \cdot Ni_{0.8} \cdot Co_{0.10} \cdot Mn_{0.10}$ will comprise approximately 33.3 grams lithium, 225 grams nickel, 28.2 grams cobalt, and 26.3 grams manganese. An entire battery pack (to empower one car) made of an estimated 960 cells would use 216 kgs of nickel, 27 kgs of cobalt, 25 kgs of manganese, and 32 kgs of lithium.

MATERIAL USED IN THE 18650 DESIGN

An individual 18650-type cell weighs 47 grams. Precise details on the structure of individual cells are guarded, though we can do some mathematical calculations to learn more.

Assuming the thickness of the cell can structure is about 0.9 mm and assuming a flat-rolled steel density of 7.8 grams per cubic centimeter, the cell can (including the top and bottom) in an 18650-type design will weigh just shy of three grams. The anode material will weigh around three grams. This means the cathode material in the cell can will weigh around 40 grams.

The stoichiometric formula of the cathode in the 18650-type design is $(Li \cdot Ni_{0.8} \cdot Co_{0.15} \cdot Al_{0.05})O_2$ means that 1 mol of Li-Ni-Co-Al material is matched by 1 mol of O_2 which weighs 32 grams. Therefore, the Li-Co-Mn material must account for the remaining eight grams of the cathode cell weight. The gram molecular weights of these components are: lithium 6.94 grams per mol, nickel 58.69 grams per mol, cobalt 58.93 grams per mol, and aluminum 26.98 grams per mol. A total of eight grams of material with a stoichiometric formula of $Li \cdot Ni_{0.8} \cdot Co_{0.15} \cdot Al_{0.05}$ will comprise approximately 0.9 grams lithium, 5.8 grams nickel, 1 gram cobalt, and 0.15 grams aluminum. An entire battery pack (to empower one car) made of an estimated 7,920 cells would use about 46 kgs of nickel, 8 kgs of cobalt, 1 kg of aluminum, and 7.1 kgs of lithium.

LFP BATTERIES

Chinese battery makers such as CATL are aware that nickel and cobalt do not occur in abundance within China. To reduce the reliance on these critical materials, Chinese battery designers have been leaning towards the prismatic battery design with a lithium-iron-phosphate (LFP) cathode design. LFP batteries are known for their cheaper manufacturing cost and modest vehicle performance.

The story of lithium-iron-phosphate (LFP) cathode material has a distinct Canadian connection. In the mid-1990s, researchers from

John Goodenough's U.K. lab proposed using LFP for battery cathodes. But their experiments showed that the material lacked the conductivity to make it suitable for vehicle batteries. Several years later, scientists at Hydro-Québec and the University of Montreal solved the conductivity problem by coating the LFP material with carbon. The batteries they assembled could not match the energy density of nickel-based cathode material, but they concluded that the use of LFP material could greatly reduce the overall cost of the battery.

In 2003, Hydro-Quebec granted a license to a tech startup company called Phostech to manufacture LFP material using a *solid-state* method, which involves mixing lithium carbonate, lithium hydroxide, and ferrous oxalate together. The mixture is ground to a fine particle size, and then heated to high temperature to fuse the compounds together. Phostech's initial efforts garnered the attention of a German firm, Süd-Chemie, who bought a controlling interest. The Germans set about using a wet method of making LFP in which a mixture of ferrous sulfate, lithium hydroxide, and phosphoric acid are dissolved in solvents, and the mixture heated to high temperature in an oven. The wet process turned out to be more expensive than the solid-state method. Phostech was eventually sold to metals refiner Johnson Matthey plc, who made little further progress.

Phostech was not alone in its LFP efforts. In 2009, the Massachusetts Institute of Technology created a spinout company called A123 Systems. The company raised money on the financial markets, but could not draw enough attention from automakers. A123 went bankrupt in 2012 and most of its assets were acquired by China's Wanxiang Group. The Chinese knew what they were doing when they acquired the assets of A123. Well before the acquisition, the Chinese were already enjoying success with LFP batteries. Many of the city buses in operation during the 2008 Beijing Olympics were powered by LFP batteries. Based on the success of this bus trial, the Chinese government decided to press ahead at a rapid clip to develop batteries with LFP cathodes.

By 2021, an estimated 40% of Chinese electric cars, with driving ranges of up to 120 kms, were using LFP batteries. Chinese battery maker, FinDreams has since made substantial improvements to the LFP battery design. An individual LFP battery cell in a Fin Dreams battery measures 90 cm x 11.8 cm x 1.3 cm and weighs 3.9 kgs. Several cells are stacked together to provide a battery pack with enough power to give a decent driving range. Chinese automaker, BYD, uses LFP batteries in its electric vehicles to provide 62 kW·hrs of battery energy; enough for a driving range of around 300 kms depending on weather conditions. These vehicles will deliver acceleration of 0 to 60 miles per hour (0-97 km/hr) in 7 seconds. The amount of lithium in a single cell is about 15% by weight. At an individual prismatic cell weight of 3.9 kgs, this implies that 0.5 kgs of lithium is used. If 100 prismatic cells are assembled into a battery pack, this suggests that 50 kgs of lithium are used. The weight of a 100-cell battery pack will be close to 390 kgs after allowing for structural mounting materials to fasten the pack to the vehicle chassis.

By comparison, a Tesla 4680-battery pack with an estimated 960 cells offers around 98 kW·hrs of battery energy; enough for a driving range of around 450 kms depending on weather conditions. A Tesla vehicle with a 4680-type battery will deliver acceleration of 0 to 60 miles per hour (0-97 km/hr) in just over 4 seconds.

The increased driving range and faster acceleration come at a cost. More lithium and more nickel are required to provide this distance and performance. The North American consumer has long enjoyed speed, acceleration, wide-open roads, and long distances The question is, will

North American consumers sacrifice these pleasures to help preserve lithium, nickel, and cobalt non-renewable resources?

Tesla doubts that the consumer will make these sacrifices. This is why Tesla continues to focus on the 18650-design and now the 4680-design. It further appears that Japanese battery makers Panasonic and Toshiba will continue, along with Tesla, to focus on the 4680 lithium-nickel-cobalt-manganese design.

But four companies think the North American consumer will make sacrifices. Israeli chemical maker ICL Group is now constructing a plant in Missouri to make LFP cathode powder material. Norway-based Freyr Battery is planning to make LFP cathode material at a factory site in Coweta County, Georgia. Utah-based American Battery Factory is planning to do likewise in Arizona. And China's Gotion High-Tech hopes to establish LFP cathode material production in Manteno, Illinois.

Despite the growing popularity of LFP batteries, they are by no means perfect. The LFP design requires careful process control at the manufacturing stage. The cathode material can readily absorb moisture from the air during the battery assembly process. If too much moisture is adsorbed by the cathode material, battery performance and longevity will be compromised.

Researchers have now determined that owing to the crystalline structure of the LFP material, electrons can move in one direction only. In the NCM architecture, the crystal structure of the cathode material allows electrons to move in two directions. The electron mobility in the NCM design is one-thousand times as fast as the electron mobility in the LFP design. Research has also suggested that after 500 charge-discharge cycles, an NCM-type battery will have 15% more capacity than an LFP battery.

The electric vehicle theme is approaching the intersection of performance and functionality. Chinese battery makers have already concluded that widespread use of cathodes using metals like nickel and cobalt will hasten global supply shortages. There are six battery makers in China that together control 56% of global battery sales. A look at their respective websites suggests they all are using the lithium iron phosphate (LFP) battery design. Korean battery makers such as LG, Samsung, and SK Innovation appear to be gravitating towards using the LFP cathode design as well. Together these Korean companies currently control 26% of the global electric vehicle battery market.

The Ford Motor Company has now made the difficult decision to use the LFP battery design in its 2024 F-150 Lightning pick-up trucks. The batteries will be designed by Chinese firm, CATL. For the acceleration reasons just described, this decision was no doubt a difficult one. Ford recognizes that the electric vehicle consumer can be segregated into two parts: those who want acceleration performance along with driving range, and those who want a dependable, affordable, electric vehicle to get them to their job and home again every day. The decision to use the LFP battery places the F-150 truck into the dependable sector instead of the performance sector. The effects on the F-150 brand image will become known soon enough.

- Lithium-ion batteries used in vehicles fall into four broad categories: the 18650-design, the 2170-design, the prismatic-design, and the 4680-design.

- The 18650, 2170, and 4680 designs use cathodes based on lithium-nickel-cobalt-aluminum (NCA) and lithium-nickel-cobalt-manganese (NCM) structures.
- The earliest cell design used in Tesla cars was the 18650-type where each cell measured 18 mm in diameter and 65 mm in length. This evolved into the 2170-design where each cell measured 21 mm in diameter and by 70 mm in length. This has now evolved into the 4680-design, where each cell measures 46 mm in diameter and by 80 mm in length.
- The prismatic design which uses lithium-iron-phosphate (LFP) cathode material is being used by Chinese battery makers. Batteries based on the LFP cathode design have poorer performance characteristics and shorter driving ranges than NCA and NCM batteries.

NEW DESIGNS. BETTER BATTERIES?

17

As battery makers contemplate the way forward from the intersection of performance and functionality, new battery designs are being explored.

THE SODIUM BATTERY

Sodium can be found on the row of the Periodic Table immediately below the row containing Lithium. The sodium atom is 25% larger than the lithium atom. In 2020, researchers at Washington State University made a battery breakthrough with a sodium-ion battery design that holds the potential to rival the lithium-ion battery design. Their research focused on constructing cathodes from sodium-iron phosphates such as: $NaFePO_4$, $Na_2Fe(P_2O_7)$, $Na_4Fe_3(PO_4)_2(P_2O_7)$, Na_2FePO_4F, and $Na_2[Fe_{0.5} Mn0.5]\cdot PO_4F$. Other more complex alloys studied included: $(O_3NaNi_{0.68}Mn_{0.22}Co_{0.10})O_2$, $Na_3V_2(PO_4)_3$, and $Na_4Co_{2.4}Mn_{0.3}Ni_{0.3}(PO_4)_2P_2O_7$.

The anode materials studied included: $Na_2Ti_3O_7$, $Na_3Ti_2(PO_4)_3$, and $Na_3V_2(PO_4)_2F_3$. This latter material showed good room-temperature battery performance. Cell capacity was shown to be about 135 m·Ah/g at an average voltage of 3.8 volts. Electrolyte compositions based on sodium salts such as $NaPF_6$, $NaN(SO_2CF_3)_2$, and $NaClO_4$ were found to be stable enough to allow for multiple charge-discharge cycles of batteries tested in the laboratory.

A battery design based on a hard-carbon anode and a cathode of composition $(O_3NaNi_{0.68}Mn_{0.22}Co_{0.10})O_2$ was found to be an optimal combination. This battery design delivered an energy density of 100–150 Wh/kg. Without question, a sodium-type battery will power a vehicle, but the issue of energy density then enters the fray. A driver wanting rapid acceleration from a standing start along with a respectable driving range will not get it with a sodium battery having energy density of 100-150 Wh/kg. A 4680-type lithium-ion battery with an energy density of 250-300 Wh/kg will however provide both acceleration and range.

The sodium battery design needs more work if it is to rival the lithium-ion design. However, the sodium design does compare favorably to Chinese batteries with LFP cathodes. The good news is, sodium-ion batteries, when fully developed, could be suitable for short-range electric vehicles. The obvious advantage offered by the sodium design is the natural abundance of sodium around the globe and its lower extraction and refining cost compared to lithium.

THE SOLID STATE BATTERY

Expect to hear more about *solid-state* lithium-ion batteries going forward. In a traditional lithium-ion battery there is a separator membrane coated with electrolyte paste between the anode and cathode cells. The anode is comprised of lithiated graphite (LiC_6). The cathode is a fused collection of metals in oxide form (nickel oxide, cobalt oxide, manganese oxide, lithium oxide).

In the solid-state battery design, the anode is made of metallic lithium, while the cathode is still of the NCA or NCM design. The separation between the cells is provided by a thin film of crystalline ceramic or solid polymer material that typically has been created by atomic layer plasma deposition of substances such as aluminum oxide (Al_2O_3). Because there is no gel or paste electrolyte in the solid-state design, safety is greatly enhanced. Because the solid-state electrolyte is a harder material, it is more impervious to lithium dendrites penetrating it. This further enhances the safety characteristics of the battery.

In the solid-state design, the battery energy density will be higher than that of a standard lithium-ion design with paste or gel electrolyte. Laboratory studies are suggesting the energy density will be higher by a multiple of up to 2.5 times. This suggests a battery that will provide a greater driving range than a standard lithium-ion battery.

Whether solid-state batteries become the norm in electric vehicles remains to be determined. A major challenge in making solid-state batteries is the advanced technology required to create the separator layer. Two publicly-listed companies pursuing solid-state designs are QuantumScape and Solid Power. As of early 2024, QuantumScape was still at the prototype stage and spending significant sums of money on research and product development. As of early 2024, Solid Power had submitted a prototype battery to automaker BMW for evaluation. This prototype uses a proprietary design of sulfide-based material for its separator.

- In 2020, researchers at Washington State University made a battery breakthrough with a sodium-ion battery design. A battery design based on a hard-carbon anode and a cathode of composition $O_{3-}NaNi_{0.68}Mn_{0.22}Co_{0.10}O_2$ was found to be an optimal combination. This battery design delivered a specific energy of 100–150 Wh/kg. While respectable, this energy level does not compare to a battery design using a cathode of $LiNi_{0.33}Mn_{0.33}Co_{0.33}O_2$ which will deliver around 250 Wh/kg.
- Sodium-ion batteries require further research. When fully developed, they could be suitable for short-range electric vehicles, but probably not for vehicles sold in the North American market.
- Expect to hear more about solid-state batteries in the near future. In the solid-state design, the separation between the cells is provided by a thin film of crystalline ceramic or solid polymer material that typically has been created by atomic layer plasma deposition of substances such as aluminum oxide (Al_2O_3). Because there is no gel or paste electrolyte in the solid state design, safety is greatly enhanced. However, a major challenge in making solid-state batteries is the advanced technology required to create the separator layer.

BATTERY METALS – WHERE FROM?

LAYERS

As long as the 18650 and 4680-type lithium-ion batteries remain in vogue, the question to consider is: where do the lithium, nickel, cobalt, and manganese materials used in battery cathode construction come from?

LAYERS

The Earth is like an onion—comprised of multiple successive layers. The outermost layer is called the *crust* and is around 30 kms thick. The chemistry of the crust is rich in silica and alumina. The layer beneath the crust is called the *mantle* and is close to 3,000 kms thick. The chemistry of the mantle is rich in iron, magnesium, and silica. The crustal and mantle rock both contain elements like copper, nickel, cobalt, manganese, and lithium.

Cobalt mineralization is present in the crustal/mantle rock at an estimated 25 parts per million (ppm). By comparison, consider that the Earth's crustal rock contains about 41,000 ppm iron and a picture of cobalt relative scarcity starts to take shape.

Nickel mineralization is present in the crustal/mantle rock at around 80 parts per million (ppm). Using the same comparator of the Earth's crust containing about 41,000 ppm iron, a picture of nickel relative scarcity starts to take shape.

The Earth's crust contains on average 20 ppm lithium mineralization which again points to relative scarcity.

The copper content of the crustal rocks is about 70 ppm. This too points to a relative lack of abundance.

Manganese is the outlier among these minerals. The Earth's crustal material contains about 1000 ppm manganese mineralization, which is still a relatively scarce amount.

PLATES

The Earth's crustal layer is broken into large pieces such that the crustal layer resembles a jig-saw puzzle. These pieces are called *tectonic plates*. At the junction between the crust and mantle is a layer of partially-melted rock. The tectonic plates are constantly in motion–albeit slow motion. Geologists reason that the tectonic plates are able to move because of the lubrication provided by this partially-melted layer of material.

The tectonic plates are not always stable. If the edge of a tectonic plate dives (subducts) beneath the edge of an adjacent plate, the friction and pressure from the grinding motion generates enough heat that the minerals in the crustal plate material melt. This molten material is called *magma*. If the resulting temperature and pressure are sufficient, extensive melting will occur and the magma will form large pools in voids in the lower part of the crust-upper mantle area. Within a magma pool, the less dense minerals will rise to the top of the pool while the denser elements will accumulate at the bottom of the pool.

The science of magma melts is complex and pivots around temperature and pressure of the molten mineral constituents. As the magma melt grows in size and pressure, the moisture contained in the melting rocks can separate from the melt and begin to circulate into and out of cracks and fissures in adjoining crustal rocks (called *country rocks*). This moisture is termed *hydrothermal fluid*. As the hydrothermal fluid circulates, it can dissolve additional mineralization from the adjoining country rocks. The magnesium, iron, and sulfur content of the country rock plays a critical role in determining the chemical composition of the magma melt. As the melt absorbs sulfur from the surrounding country rock, copper, cobalt, and nickel sulfide mineral structures form.

Not all tectonic plates were created equal in their chemical composition around the globe. Across the annals of geological time, selected tectonic plate subduction zones around the globe developed elevated concentrations of minerals such as cobalt, nickel, lithium, copper, and manganese in the magma. Today, it is these locations that the mining industry has focused on to economically mine, extract, and refine these minerals.

Geologists have arrived at several different complex models that seek to explain how temperature, pressure, density, and hydrothermal fluids have all influenced the deposition of cobalt, nickel, lithium, copper, and manganese mineralization in the crustal layer.

For example, copper has a density of 8.9 g/cm³ and a melting point of 1085°C. Manganese has a density of near 7.2 g/cm³ and a melting point of 1246°C. In a magma melt, the copper will melt before the manganese. Nickel has a melting point of 1450°C and a density of 8.9 g/cm³. Cobalt has similar properties. Iron has a higher melting point at 1538°C and a density of 7.86 g/cm³. Because of the density differences between these mineral constituents, molten manganese mineralization will rise to the top of the magma pool. Copper, nickel and cobalt will intermix in a layer beneath the manganese. Iron will melt last, but owing to its density will intermix with the manganese.

One geological model suggests that as these intermixed layers combined with hydrothermal fluids, the molten material oozed up to the surface of the crustal material where it cooled and solidified. Another model suggests that as the hydrothermal fluids worked their way up from depth, the iron content associated with the molten material separated off as a result of changes in the acidity (pH) of the fluids and remained behind. The mineral-rich fluids reaching the surface of the crustal layer cooled and solidified. Another model, particular to manganese, suggests that some 2.5 billion years ago crustal layer rock formations enriched in manganese mineralization came to be submerged in ancient salt water seabed deposits. Over geologic time, these sedimentary seabeds were transformed to dry land through repeated uplift and movement of the tectonic plates.

If these models seem calm, be assured that not all magma events are so. In some cases, as the pressure builds in the magma chamber, the melted material will seek to relieve itself of the pressure buildup by squeezing upwards into joints between the tectonic plates. If the pressure is high enough, the magma will explode upwards as a volcanic eruption. The veins of magma that ultimately do not make it all the way to surface will eventually solidify. These volcanic locations with underground solidified mineralized zones are prime targets for the mining industry to exploit around the globe.

The magma material from a volcanic event that arrives at the surface of the crustal layer can be exposed to the forces of weathering and erosion over vast intervals of geologic time. In sub-tropical and tropical climates, where the magma was enriched in iron and nickel, this weathering has left behind concentrations of iron and nickel mineralization called *laterite formations*.

The complexities of tectonic plates, density, temperature, and pressure are one matter. The other issue is that when the mineralization arrived at the surface layer of the crust, it did not do so in a uniform manner. Today, certain countries around the world are blessed with an abundance of mineralization thanks to the unevenness of geological events millions of years ago. For example, Mexico, Brazil, South Africa, Russia, and northwest Australia host manganese deposits. Countries such as Cuba, the Philippines, and Indonesia host vast economic deposits of nickel. African countries like the Democratic Republic of Congo now command huge geopolitical leverage thanks to their cobalt deposits. The five prime locations in the world for copper deposits are: Chile, Peru, Mexico, the south-western U.S., and south-Australia. While copper is not used in battery construction, electric vehicles have up to four times the amount of copper wiring than do gasoline-powered vehicles.

As political policy-makers press ahead with their agenda of electric vehicles, expect geopolitical tensions to come to the fore. However, the issue that is even more critical to assess is whether the Earth has enough of these battery minerals to fully pursue an electric vehicle agenda around the globe.

- The Earth's crustal layer is broken into large pieces such that resemble the pieces of a jig-saw puzzle. These pieces are called *tectonic plates.*
- If the edge of a tectonic plate dives (subducts) beneath the edge of an adjacent plate, the friction and pressure from the grinding motion generates enough heat that the minerals in the crustal plate material melt. This molten material is called magma.
- As the magma melt grows in size and pressure, the moisture contained in the melting rocks can separate from the melt and begin to circulate into and out of cracks and fissures in adjoining crustal rocks.
- Today, certain countries around the world are blessed with an abundance of mineralization thanks to the unevenness of geological events millions of years ago. Depending on the political leadership in these counties, the pursuit of mineralization can trigger geopolitical tensions.

BATTERY METALS - WILL WE HAVE ENOUGH?

The metals and minerals used in battery making are a finite, non-renewable resource. When a finite resource is gone, it is gone for good.

The International Energy Agency (IEA) reports that in 2022 there were 10.5 million electric vehicles sold globally. The website *www.evvolumes.com* (owned by data analytics firm J.D. Power) breaks this number down. Battery powered electric vehicle sales in 2022 amounted to 7.7 million units. Hybrid sales amounted to 2.8 million units. Of the 7.7 million units sold in 2022, 6.7 million of those were sold in China. The 2023 numbers are pointing to total global sales of 14 million units with 4 million of these being hybrid vehicles.

The burning question is: are we making a mistake by so urgently rushing towards the use of electric vehicles? The metals used in battery manufacture also enjoy other critical, high-tech uses in our society. Is there enough of these battery metals to complete the transition to an electric vehicle society while still having enough left over to satisfy the other high-tech applications for future generations? Let's approach this issue by first looking in detail at the metals that are used to make vehicle battery cathodes.

The following examination hinges around two key definitions: *reserves* and *resources*. A reserve of a given mineral is a quantity that geologists have proven exists and that can be mined and processed economically. A resource of a given mineral is a lower-grade quantity that on the basis of geological evidence has reasonable prospects for eventual extraction. The terms *reasonable* and *eventual* are both judgement calls by geological professionals. The critical issues in extracting material that has been labelled a resource are: technology, ore grade, and processing cost. Does the technology exist to crush, grind, and recover lower grades of ore? How does the overall cost of extraction and recovery compare to current prices for the material in question?

COBALT

Cobalt was recognized in 1735 by Swedish chemist Georg Brandt when he chemically separated a mineral substance from a sample of unusual-looking ore obtained from a Swedish copper mine. The name cobalt comes from the German word, *kobold*, meaning *evil spirit*. At the time of Brandt's discovery, miners in Sweden were suffering premature deaths from arsenic contained in the copper ore. Brandt initially thought the substance he had separated was related to arsenic, hence the choice of a name meaning *evil spirit*.

One geographic location richly endowed with cobalt mineralization is what we today recognize as part of Zambia and the Democratic Republic of the Congo (DRC). Nearly 70% of the world's cobalt is sourced from this region.

Other locations of note around the globe with cobalt occurrences include Noril'sk-Talnakh and Pechenga in Russia, Voisey's Bay in Canada, the Kambalda region of western Australia, and the Bushveld region of South Africa. Cobalt mineralization in all of these areas is present along with copper and nickel mineralization. The cobalt is separated from the copper and nickel mineralization during mineral processing.

China is adept at making new geopolitical friends. Chinese metal refiners have wasted little time in making political friends in the DRC. In a clever bit of cheque-book diplomacy, to the tune of $9 billion, China inked a joint venture with the DRC government. China obtained the rights to the vast copper and cobalt resources of the North Kivu region in exchange for providing infrastructure including roads, dams, hospitals, schools, and railway links. Thanks to this strategic maneuvering, China controls about 85% of global cobalt supply.

At first glance, this appears to be a strange move given that the Chinese battery makers are using the LFP design of battery which does not contain cobalt. But a deeper look suggests this move to make political friends in the DRC is a bold geopolitical gambit designed to ensure that other countries do not have ready access to cobalt. This gambit was motivated by the fact that cobalt is used for a variety of applications including: catalysts for chemical processes; pigments for glass, enamels, pottery and ceramic manufacture; magnets; radio-isotopes for medical treatment; superalloys for jet engine turbine blades, alloys for corrosion-resistant industrial applications, alloys for surgical instruments and prosthetics; and electronic connectors on integrated circuit boards. But the big use of cobalt is to make the small lithium-ion batteries that are essential for smart phones and laptop computers. Control a major source of cobalt and you control smart phones and laptop computers—an important segment of the global economy. Control the cobalt market and you have significant influence over any company seeking to use cobalt, whether it be for battery cathodes, turbine blades, or surgical instruments.

The mining, processing, and refining of cobalt-bearing rock is energy consuming and polluting. The processing first involves crushing and grinding the mined ore. The crushed material is then roasted at temperatures of around 1300°C to produce what is called a *matte* material. The matte is then subjected to an elevated temperature treatment to further remove unwanted impurity elements from the cobalt mineralization. Refining is next required to further isolate the cobalt content. In the refinery, the matte material is leached in solutions containing hydrochloric acid, ammonia solution, or sulfuric acid. This is followed by solvent extraction to precipitate out the cobalt mineralization. The recovered material is then subjected to electro-chemical treatment to obtain cobalt with sufficient purity to be used in alloy applications and battery applications.

A 2021 article in the journal *International Journal of Life Cycle Assessment* detailed the Chinese cobalt refining process and its negative effect on air, water, and soil. The article stresses that electro-chemical refining of cobalt is an energy intensive process that uses power very likely generated at coal-fired generating plants. The gently worded conclusion was: chemical inputs and electricity consumption are primary sources of potential environmental impact in China's cobalt production.

The United States Geological Survey (USGS) estimates the global reserves of cobalt-containing ore to be 7.6 million tons (6.9 million tonnes). Of this figure, half is contained in geopolitical hotspots like the DRC and Zambia. The USGS notes that for 2022, 70% of global cobalt supply came as a by-product of copper mining in the DRC. Fully 80% of this cobalt ended up in the hands of Chinese metal refiners.

Overall recovery efficiency of cobalt varies between 30-70% depending on the initial concentration of cobalt in the mined ore and how tightly bound the cobalt is in the ore mineral structure. Generally, cobalt will occur in sulfide-type mineral structures. This means that the waste tailings from a mineral processing operation will have a sulfide content. As these tailings are exposed to moisture, acid compounds form which are deleterious to the surrounding environment. Cobalt mineral processing is also responsible for emissions; a report from consulting firm Roskill, states that cobalt refining in 2021 was responsible for 1.6 million tonnes of carbon dioxide emissions.

Recovering cobalt, or any other mineral, as a by-product in a mining operation is never 100% efficient. Assume that globally, the 7.6 million tons (6.9 million tonnes) of cobalt reserves referenced by the USGS can be recovered into matte material at an efficiency of 60%. Assume that further downstream refining to purify the cobalt is about 80% efficient. This means there is roughly 3.3 million tonnes of cobalt metal recoverable from the 6.9 million tonne reserve figure. The website *www.mining.com* (owned by Canadian media firm Glacier Media) suggests that 175,000 tonnes of cobalt metal is consumed each year for electronics, and specialty alloy applications. If this usage rate holds level, this means that **by about 2043 the globe will be depleted of its cobalt reserves. And this bit of math does *not* include the cobalt used in electric vehicle batteries.**

Peak Cobalt?

In a previous chapter, it was estimated that a 4680-battery with 960 cells will contain 27 kgs of cobalt. Between 2023 and 2030, forecasts shown on the *www.evvolumes.com* website suggest that nearly 186 million electric passenger vehicles will be made in worldwide by 2030. Estimates show that 35 million hybrid vehicles will be made. Assuming that 40% of these non-hybrid vehicles will use batteries of the 4680 design containing cathodes made partly of cobalt, the amount of cobalt to be consumed will be about 286,000 tonnes per year. Add this to the 175,00 tonnes per year used for all other cobalt applications and the calculations show that **if electric vehicle sales projections hold true, by 2030, barring a major new cobalt discovery, the planet will be largely exhausted of its cobalt reserves.** The Chinese have evidently already done these calculations and have been motivated to boldly moved to gain control of the cobalt industry in Africa. We have all heard the arguments for peak oil. We could soon be hearing arguments for peak cobalt. This will drive prices higher for not only electric vehicle batteries but for electronic goods and specialty alloys that contain cobalt. The beneficiary of these higher prices will be China. The Chinese geopolitical tentacles attached to cobalt are *concerning* to say the least.

The USGS has cited a global cobalt resource figure of 25 million tonnes. However, this figure has not been supported with an indication of ore grade or extraction cost estimates. To assume that these resources can all be recovered to support industrial cobalt consumption plus battery making, is short-sighted and naïve.

NICKEL

In 1751, Swedish scientist Axel Fredrik Cronstedt was tasked with studying a piece of arsenic-containing rock from a copper mine in the Hälsingland region of central Sweden. His research revealed a new metal which he named nickel.

Sulfide Ores
Magma that absorbed sulfur from the surrounding country rock has given rise to what geologists refer to as *sulfide nickel* ore deposits. Sulfide nickel occurrences are generally large, lower-grade deposits

containing grades less than 1% nickel. Examples include Noril'sk-Talnakh in Russia, western Australia, the Brazilian state of Goiás, and the Canadian regions of Thompson (Manitoba), Sudbury (Ontario), and Voisey's Bay (Labrador).

Despite the lower grades of sulfide nickel deposits, their extractive metallurgy lends itself to economic recoveries. The bad news is, these larger deposits are becoming depleted. Russia has a reported seven million tonnes of nickel reserves remaining. Given the geopolitical fallout from the 2022 Ukraine invasion, this nickel may well remain stranded if Russia remains hobbled by economic sanctions for years to come. Canada has a reported two million tonnes of nickel reserves remaining. By 2030, the amount of nickel sulfide ore remining to be mined in Canada could be insignificant barring a major new discovery. The two countries that have nickel sulfide reserves remaining are Australia with a reported seven million tonnes and Brazil with a reported 16 million tonnes.

Nickel in sulfide deposits is present in a mineral structure called *pentlandite*. In order for nickel sulfide mineral to be used in battery cathode manufacture, the sulfide structure must be converted to nickel sulfate format. To accomplish this, the mined sulfide ore is crushed and ground into a coarse powder format using rock crushers and grinding mills. The coarse material is then subjected to a flotation process. Picture a big bathtub with air bubbling through it. The coarse powder is mixed with chemical reagents and added to the bathtub. As the air bubbles rise to the top of the water, they carry with them nickel-iron-sulfide particulates which are skimmed off and collected for further processing. The skimmed-off material is then treated with either acid or ammonia to further liberate the nickel. This chemically treated material is then smelted in a kiln at near 1300°C to cause the iron mineralization to separate off as molten slag. The remaining material is called *nickel matte* and comprises 70-75% nickel. The matte material is subjected to high pressure acid leaching to dissolve it into solution. The solution containing the nickel is then subjected to an electrowinning process in which an applied electric current causes nickel to precipitate out of solution and accumulate at a cathode at a purity of near 99.9%. This highly refined material is then subjected to treatment with sulfuric acid and oxygen to create nickel sulphate which is what battery makers need to construct lithium-ion battery cathodes. The International Energy Agency (IEA) suggests that processing of sulfide nickel ore generates 10 tonnes of carbon dioxide every time one tonne of nickel ore is processed.

Laterite Ores

Magma with a nickel and iron content that is brought to the surface has the potential to be eroded by weathering forces over millions of years. The weathering removes some of the iron content and leaves much of the nickel content intact. Weathered deposits are what geologists refer to as *laterite deposits*.

The typical ore grade of a laterite occurrence is 1-1.5% nickel. Laterite type deposits account for near 70% of the known nickel occurrences in the world and are located in tropical and sub-tropical regions of the globe. For example, Indonesia hosts substantial lateritic geological deposits as does the Pacific island of New Caledonia.

The mineralization of these laterite deposits takes the form of ore types such as *limonite* and *saprolite*. The molecular weights of these mineral structures are such that they do not lend themselves to a flotation process during mineral processing. Moreover, lateritic ore has a moisture content associated with it. In order to remove the moisture, the ore must be heated in large kiln furnace. The energy consumed is substantial.

M.G. BUCHOLTZ

Saprolite ore has a higher nickel content and lower iron content that limonite ore. Kilned saprolite ore lends itself favorably to a smelting process in electric arc furnaces, which further consume significant amounts of energy. The end-product is a ferro-nickel compound, sometimes called *nickel pig iron*.

The ferro-nickel from laterite deposits is often used in cheaper expressions of ferritic stainless steel. My kitchen sink was sold to me at Home Depot as being "stainless steel." Judging from the price I paid, I knew my sink was not the same type of stainless steel as my expensive cookware is made from. However, technically the sink was stainless steel, so there was no misleading advertising involved. Four years later, the light brown stains on my sink confirm the presence of iron in the alloy. My sink is made of stainless steel that came from a saprolite lateritic nickel deposit likely somewhere in Indonesia.

Limonite ore has a lower nickel content and higher iron content than saprolite and as such does not economically lend itself to smelting. Instead, kilned limonite ore is given an energy-intensive roasting treatment at near 750°C to remove oxygen molecules from the limonite mineral structure. The roasted material is then processed by a high-pressure hydrogen reduction treatment (200°C and 400 psi pressure in autoclave cooking ovens) to separate off some of the nickel mineralization. The remining leachate material is treated with hydrogen sulfide to further precipitate out the remainder of the nickel content. This residual material is called *mixed hydroxide precipitate* (MHP), or *mixed sulfide precipitate* (MSP). The International Energy Agency (IEA) suggests that processing of nickel laterite ore generates almost six times *more* carbon dioxide than nickel sulfide ore processing.

Chinese and Indonesian scientists are now claiming they can produce nickel sulphate from leached MHP material with sufficient purity for battery cathode manufacture. However, the chemical conversion process creates a significant amount of waste effluent. Until late 2020, Indonesian nickel processors were dumping the effluent waste into the ocean. In early 2021, the Indonesian government ordered a halt to the issuance of permits for this disposal process. Where the effluent is being disposed of now is uncertain.

Here again, there is a disconnect. Chinese battery makers are mainly using the lithium-iron-phosphate (LFP) design of battery which contains no nickel in the cathode. The forward-thinking Chinese were quick to realize that nickel is an important global metal. They were quick to realize that nickel laterite deposits represent the future of nickel mining and that over 90% of all the nickel produced in the world is used for making stainless steel, specialty nickel-based alloys, and for electroplating. Following on their success in the DRC, they realized that it would be beneficial to establish friendly relations with a nearby country that produces nickel. In 2019, Chinese metal refiners with financial help from the Chinese government, executed a cheque-book diplomacy strategy by approaching political leaders in Indonesia and offering to invest $15 billion in the economy. Indonesian leaders accepted the proposal and moved up the date for a planned ban on raw nickel ore exports by Indonesian producers. This started global nickel prices on an upward trajectory and made nickel users nervous as they scrambled to find new sources. To get around Indonesia's planned export ban, the Chinese established their own nickel processing and refining operations in Indonesia.

The United States Geological Survey (USGS) estimates the global reserves of nickel-containing ore to be 100 million tonnes. Recovering nickel, or any other mineral, in a mining operation is never 100% efficient. Assume that globally, the 100 million tonnes of nickel reserves referenced by the USGS can be recovered at an efficiency of

90%. Assume that further downstream refining to purify the nickel is about 85% efficient. This means there is roughly 76 million tonnes of nickel metal recoverable from the USGS reserve figure.

The global economy uses about three million tonnes of nickel per year for all applications. If this usage rate holds level, this means that **in about 30 years the globe will be depleted of its nickel reserves. While this timeframe does not raise any immediate alarms, this bit of math does *not* include the nickel used in electric vehicle batteries.**

Peak Nickel?

In a previous chapter, it was estimated that a 4680-battery with 960 cells will contain 216 kgs of nickel. Between 2023 and 2030, forecasts shown on the *www.evvolumes.com* website suggest that nearly 186 million electric passenger vehicles will be made by 2030 worldwide. Estimates show that 35 million hybrid vehicles will be made. Assuming that 40% of these non-hybrid vehicles will use batteries of the 4680 design containing cathodes made largely of nickel, the amount of nickel to be consumed will be about 16 million tonnes per year. Add this to the three million tonnes per year used for all other nickel applications and the calculations show that **if electric vehicle sales projections hold true, the planet is quickly marching towards a scenario where economically extractable nickel reserves are facing exhaustion.**

The Chinese have already done these calculations. This is why they so boldly moved to make political friends in Indonesia. The arguments for peak cobalt might soon be echoed in the form of arguments for peak nickel. Unless, of course, North American battery makers resort to making batteries that do not rely on nickel in the cathode design. However, this would result in battery-powered vehicles with shorter driving ranges and reduced acceleration and highway performance–factors the North American consumer would be loath to accept.

The geopolitics of nickel might intensify towards 2027. In August 2022, the U.S. government signed the *Inflation Reduction Act* (IRA) into law. Electric car buyers will be eligible for tax credits providing 40% of the vehicle's battery materials and components are extracted or processed in the U.S. or in a country that has a free trade agreement with the U.S. This manufacturing threshold will increase annually, and by 2027, 80% of the battery must be produced in the U.S. or a partner country to qualify for tax credits. The U.S. government will have to make haste to sign a free trade agreement with nickel producing nations like Brazil and Australia in order to meet the 2027 deadline.

The USGS cites a global nickel resource figure of 300 million tonnes. Breaking this figure into laterite and sulfide geology shows 60% of this resource figure to be laterite occurrences. Nickel metal obtained from laterite ores is not necessarily suitable for all nickel alloy end uses. As well, the pollution factor from processing laterite ores is an estimated six times that of the pollution factor from processing sulfide ores. Moreover, this USGS resource figure broadly includes resources with mineral grades of greater than 0.5% nickel and has not been supported with an indication of ore-extraction cost estimates. Low ore grades further add to the energy intensity, capital cost, technical challenges, and environmental operating risks of mining and processing.

MANGANESE

The discovery of manganese also has a Swedish connection. In 1774, Swedish chemist Carl Wilhelm Scheele speculated that a dark-colored ore found by miners was a new mineral. Later that year, chemist Johan Gottlieb Gahn melted a sample of the black mineral substance in the presence of charcoal. The resulting metal was given the name *manganese*.

Glacier Media website *investingnews.com* estimates the global reserves of manganese-containing ore to be 690 million tonnes. This figure has been compiled from various geological literature sources. Over 85% of the global manganese resources are found in Mexico, South Africa, and Ukraine. Other smaller manganese occurrences are found in China, Australia, Brazil, Gabon, and India.

Manganese is widely used in industry. Its uses include: dry cell disposable batteries, paint pigment, fungicides, animal feed, glassmaking, chemical process catalysts, ferro-manganese additions in steelmaking, specialty steel alloys, and specialty aluminum alloys. The steel industry alone accounts for nearly 90% of manganese consumption each year.

Processing of manganese ore to a level of purity suitable for use in battery cathodes is a multi-step process. The mined ore is crushed and ground to a fine texture before being subjected to magnetic separation to remove impurities. The material collected from the magnetic separation is acid leached to remove the iron content. The remaining manganese-rich material is subjected to electrolytic separation to obtain what the battery industry calls *high purity electrolytic manganese metal* (HPEMM).

Assume that globally, the 690 million tonnes of manganese reserves can be extracted at an economic efficiency of 80%. Assume that refining of the extracted ore is about 85% efficient. This means there is near 470 million tonnes of manganese metal recoverable in the world. This figure does not raise any immediate alarms; however, it does not include the manganese used in electric vehicle batteries.

Peak Manganese?
In a previous chapter, it was estimated that a 4680-battery with 960 cells will contain 25 kgs of manganese. Between 2023 and 2030, forecasts

shown on the *www.evvolumes.com* website suggest that nearly 186 million electric passenger vehicles will be made by 2030 worldwide. Estimates show that 35 million hybrid vehicles will be made. Assuming that 40% of these non-hybrid vehicles will use batteries of the 4680 design containing cathodes made partly of manganese, the amount of manganese to be consumed will be about 1.85 million tonnes per year. Annually the global consumption of manganese is just over 25 million tonnes. **Factor in the manganese usage for electric vehicle batteries and as the year 2040 draws near, the planet will be largely exhausted itself of its manganese reserves.** Unless, of course, North American battery makers resort to making batteries that do not rely on manganese in the cathode design.

COPPER

Copper is not specifically used in battery construction. However, a battery powered vehicle contains a significant amount of copper wiring. The United States Geological Survey (USGS) suggests that the amount of copper mineralization in reserve around the globe is 870 million tonnes.

Peak Copper?
The USGS suggests this figure must be reduced to account for those parts of the globe that are prone to geopolitical tension. This leaves an estimated 710 million tonnes. Assume that the mining recovery efforts will be 80% efficient and that refining efforts will be 85% efficient. This suggests that there is in the range of 482 million tonnes of copper reserves known and available. Annual global consumption of copper is close to 28 million tonnes. This implies that **the global reserves of copper will be facing a depletion scenario in about 20 years.** A typical electric vehicle will use up to 80 kgs of copper wiring. Between now and 2030, if 186 million electric vehicles are manufactured as

sales projections indicate, a copper usage figure of 15 million tonnes to make these vehicles is a reasonable estimate. Critics of this copper argument will point to the significant quantity of copper resources that the USGS estimates to be recoverable. The problem is that at present, the average grade of copper ore being mined in the world is less than 0.5% which equates to less than 5 kilograms of copper mineralization per 1000 kilograms of ore extracted. The market price of refined copper is in the $8 per kilogram range. This implies that one tonne (1000 kgs) of copper ore must be able to be mined and refined economically for less than $8 per kilogram to provide the mine owner with a profit on capital invested.

As time marches on and as the grades of the remaining copper ore reserves decline, the price of copper has but one way to move– upwards. This will cause not only the price of electric vehicles to increase but also the price of all consumer goods that rely on copper for wiring or circuitry. **The effects on the entire global economy will be notable.** Low ore grades further add to the energy intensity, capital cost, technical challenges, and environmental operating risks of mining projects.

LITHIUM

Hundreds of millions of years ago in a few select locations around the globe, volcanic activity forced lithium-enriched magma to squeeze into cracks and voids in rock structures situated closer to the surface of the Earth's crustal layer. Geologists refer to these concentrated intrusive formations as *pegmatite dikes* or *hard rock formations*.

The naming of lithium also has a Swedish connection. In the late 1700s, a unique looking mineral was discovered on the Swedish island of Utö. In 1817, chemist Johan August Arfvedson of Stockholm finally analyzed the rock sample and deduced that it contained a previously unknown metal, which he called *lithium*. The name lithium is taken from the Greek word *lithos*—meaning stone, a reference to its hard-rock presence.

Adding lithium to glass reduces the energy production costs by lowering the melting temperature and the viscosity of the melt. This feature is what allows for better quality windshields to be made for the automotive industry and for more temperature-tolerant spacecraft heat shields. Lithium addition to ceramics reduces the cost of production by lowering the kiln firing temperature needed which means less energy consumption. Lithium added to ceramics also lowers the thermal expansion factor of the ceramic. A lower thermal expansion factor allows a ceramic dish to be removed from a hot oven into a room temperature environment without the risk of the ceramic container breaking.

Over 75% of lubricating greases sold globally are based on a lithium composition, which allows the greases to perform effectively at a wide range of temperatures and moisture conditions. The aluminum industry adds lithium chemicals to high temperature molten aluminum alloys which lowers the melting temperature and viscosity of the melt which translates into reduced energy consumption. Lithium products can also be used for air treatment. For example, lithium hydroxide is used to absorb carbon dioxide from the air in confined spaces such as submarines and the International Space Station.

Hard Rock Lithium
One prolific area of hard rock lithium concentration is in western Australia. The Wodgina deposit contains an estimated 259 million tonnes of lithium mineralization grading 1.17%. The nearby Pilbara deposit contains an estimated 309 million tonnes of lithium

mineralization grading 1.14%. The Greenbushes deposit located 250 kms south of Perth is the largest hard rock lithium deposit in the world with an estimated 360 million tonnes of lithium mineralization grading 1.5%.

Of course, the Chinese have been watching the lithium mining sector in Australia and taking action. Tianqi Lithium now owns a 51% controlling interest in the Greenbushes deposit. How or why the Australian government allowed this to happen is a question that many Australians no doubt want answered.

A smaller lithium deposit in Sonora, Mexico containing about 37 million tonnes of mineralization grading 4% further illustrates the aggression of the Chinese. While this deposit is relatively small, it attracted the Chinese company, Ganfeng Lithium, which now owns 100% of it. In early 2023, the Mexican government made overtures about nationalizing the Mexican lithium industry. No doubt some cheque-book diplomacy will be orchestrated to ensure the project remains under Chinese control.

Over the past two years, Brazil has made several lithium hard rock discoveries and now is becoming a significant lithium player. Several companies with a Canadian connection and at least one with an Australian connection have been focusing on Minas Gerais state. The reserves identified so far are just over 100 million tonnes of ore grading 1.3%.

Zimbabwe is another area that has potential for hard rock lithium potential. However, the mere mention of Zimbabwe invokes images of geopolitical intrigue. True to form, the Chinese are already making inroads into the country. As of mid-2023, the Chinese mining firm, Huayou Cobalt, had started trial production at its Arcadia lithium project thought to contain many tens of millions of tonnes of lithium ore.

Hard Rock Processing

In a hard rock mining scenario, the mined rock is crushed and ground into a finer particle size. The ground particles are then passed through a kiln and heated to about 1100°C. The intense heat energy causes the lithium crystal structure to alter from a dense α-spodumene monoclinic structure to a less dense mixture of the tetragonal β-spodumene and g-spodumene structures. The alteration causes the particles to develop fine cracks. After exiting the kiln, the calcined, structurally-altered material is re-heated to 250°C in the presence of sulfuric acid. The acid penetrates the fine cracks in the rock particles and leaches out lithium in the form of lithium sulfate. The sulfate material is then mixed with sodium carbonate to produce lithium carbonate and a sodium sulfate by-product. The lithium carbonate material is next mixed with calcium hydroxide (lime) to produce lithium hydroxide.

A major drawback of this process is the generation of a sodium sulfate waste by-product and a calcium carbonate sludge product, both of which require disposal. Improper disposal practices can lead to environmental damage. The other concern with this process is the amount of hydrocarbon energy used in the heating process.

No matter what the design of the battery in an electric vehicle, lithium is required. The mining and processing of lithium-containing rock requires significant fossil fuel hydrocarbon energy.

Consumers who think that the lithium-ion batteries in electric vehicles are environmentally friendly may wish to rethink that notion.

Clay Lithium Deposits

Lithium also occurs in clay deposits. The area along the border between Nevada and Oregon in the U.S. is garnering attention with its lithium-bearing clay deposits. An estimated 15 million years ago, this area was the site of numerous volcanic eruptions that deposited thick layers of ash on the terrain. Over the hundreds of years that followed, hot hydrothermal fluids oozing from the sites of the old volcanic calderas caused the composition of the ash to transform into clay. Geologists think that the hydrothermal fluids leached magnesium ions out of the clay. Lithium ions then assumed the place of the magnesium ions in the clay mineralogical structure. The lithium content of these clay formations is usually well under 1%, but if sufficient tonnage exists, a mining extraction scenario can be economic.

As of mid-2023, Canadian-headquartered Lithium Americas was advancing its Thacker Pass project area along the Nevada-Oregon border towards production. Its calculations show a resource of 13 million tonnes of clay material grading 0.22% lithium. So desperate was General Motors for a future supply of lithium for batteries that it offered $650 million to Lithium Americas for an ownership stake in the project, which as of early 2024 has not yet been advanced to the mining stage. Further south, in Clayton Valley, Nevada, Canadian company Century Lithium is engaged in a feasibility study on another clay project to determine whether it can successfully extract lithium carbonate mineralization from the clay material. Drill results suggest the project area could contain 200 million tonnes of clay, grading 0.11% lithium. A capital cost of $500 million will be needed to advance the project to a production scenario. No doubt other automakers are already eyeing this project.

Clay Deposit Processing

As of late 2023, lithium clay deposits are largely untested at scale. At the Lithium Americas clay-based lithium project along the Nevada-Oregon border, the extracted clay material will be leached with sulfuric acid to separate the clay from the lithium mineralization. The clay particulates in solution will then be filtered off. The acidity of the remaining solution will be neutralized by adding crushed limestone before being exposed to heat to precipitate out the magnesium content in the form of magnesium sulfate. Soda ash (sodium carbonate) will then be added to the remaining solution to produce lithium carbonate solids. These solids will then have to be sent to a refiner for conversion to lithium hydroxide monohydrate suitable for cathode manufacture. According to a company technical report, laboratory work has proven this process to be 83% efficient at extracting lithium from the clay material. More details will emerge in 2027 when mining operations are slated to begin. The upfront capital to build the mine site is being provided through a $2.26 billion dollar loan from the U.S. government.

Salar Lithium

Lithium also occurs in brine deposits a few meters beneath the surface of the terrain. The part of the globe that is the focus of brine deposits is the *Lithium Triangle*, the area where the countries of Argentina, Chile, Bolivia, and Peru all come together. These brine deposits (called *salars*, a Spanish expression meaning salt flat) are estimated to contain nearly 66% of global lithium resources. The USGS notes that this 400,000 square kilometer area (150,000 square miles) contains over 100 salars, ranging in size from a few square kilometers to over 9,000 square kilometers.

Geologists estimate that some 30 to 40 million years ago, the two tectonic plates that define the west coast of South America (Nazca plate and the South American plate) underwent a period of subduction with the Nazca plate slipping beneath the South American plate. This subduction resulted in intense volcanic activity and crustal uplift. Ancient shallow lakes became stranded by this uplift to form basins. Lithium, sodium, sulfate, carbonate, potassium, and magnesium sediments leaching from the surrounding volcanic ash have over the ensuing millions of years amassed at these ancient lakes. Over time, wind erosion and blowing dirt caused these lakes to become covered with silt and dry out. What remains today are the salar formations.

The largest of these salars is Salar de Uyuni located in Bolivia. The U.S. Geological Survey estimates that the various salars in Bolivia collectively hold some 21 million tonnes of brine material. Up until mid-2023, the Bolivian government maintained tight control over its natural resources and their exploitation. But then suddenly that policy vanished. With the stroke of a pen, Russian state nuclear company Rosatom and China's Citic Guoan Group inked a deal to exploit Bolivia's lithium. How western governments allowed the Russians to make this investment in light of the Ukraine situation is a question that remains unanswered. Russia will invest $600 million to advance the deposits towards production. The Chinese will invest $857 million and will go so far as to consider establishing a battery plant and perhaps even a vehicle assembly plant in Bolivia. What these partners seem to have overlooked is the fact that the Bolivian salar brines have high levels of magnesium, boron, and sulphate mineralization. The estimated magnesium to lithium ratio is 18:1 which is more than double the ratio of salar brines in Argentina and over six times the ratio of Chilean brines. The recovery of lithium from the Salar de Uyuni at a purity suitable for battery-making will pose a significant engineering and financial challenge.

The area around Atacama, Chile has long been recognized for its lithium resources. With an estimated resource of 6.3 million tonnes grading 0.14%, the Atacama salars have a magnesium to lithium ratio of around 3:1. This attractive chemistry is why Chilean-based Sociedad Quimica y Minera de Chile S.A. (NYSE:SQM) and U.S.-based Albemarle (NYSE:ALB) are actively exploiting the lithium resources in the area. Material extracted from this salar currently supplies nearly 40% of the lithium for the U.S. market and 85% of the lithium for the E.U. market.

Salar Lithium Processing
Images of the white-colored salars evoke thoughts of cleanliness and thoughts of saving the environment. Such is not necessarily the case.

Mineral-rich salar fluids are pumped to the surface through shallow bore holes drilled into the salar formations. The salar fluids are pumped into a sequential, cascading array of settling ponds. Owing to the elevation of the area (8,500 feet above sea level) and the latitude (23 degrees south of the equator), there is a significant amount of solar radiation that impacts the area. After about an 18-month timeframe, a significant amount of water has evaporated from the settling ponds and the salar fluids are now even more concentrated in mineralization. The fluids are collected and trucked to nearby processing plants where lithium carbonate and lithium hydroxide are extracted.

The residual material remaining in the settling ponds is comprised of magnesium, potassium, sodium, and other minerals that are of little economic use. This material is cleaned from the settling ponds and disposed of in waste piles. The processing plants to which the concentrated brine has been delivered use fresh water pumped in from the nearby mountain ranges to dilute the concentrated brine. The diluted solution is then passed through selective filtration steps to remove suspended solid impurities such as magnesium and calcium. The filtered brine is then treated with sodium carbonate at 90–95°C to precipitate the lithium content as lithium carbonate. The lithium carbonate material is then further filtered to remove remaining contaminants. The carbonate material is then reacted with lime $(Ca(OH)_2)$ to obtain lithium hydroxide monohydrate which is the desired material for battery cathode manufacture. This overall process is far from efficient. One problem with the process is that lime $(Ca(OH)_2)$ has a low solubility which means some lithium ions are lost in the solid waste stream resulting from the process. Large lithium producers such as SQM and Albermarle do not readily disclose the efficiency of their processes. The poor efficiency of the process is hinted at because the global pricing of lithium hydroxide battery grade material is based on a minimum purity of 56%; if the process was more efficient, the global market would be structured around a higher minimum purity level.

To add some numbers to this efficiency scenario, consider that a brine deposit containing 1,400 ppm lithium (typical of the Salar de Atacama in Chile) will have to have nearly 200,000 liters (240,000 kgs by weight) of brine material extracted and processed in order to recover 1,000 kgs of lithium carbonate.

The amount of water lost to evaporation in the settling ponds is estimated to be near 100,000 liters per tonne of lithium carbonate material obtained. The longer-term impacts of salar mineral extraction have not completely been quantified, given the desire of operating companies like SQM and Albermarle to safeguard their operational details from environmental scrutiny. Hydrological records in the Lithium Triangle remain either conveniently unavailable or incomplete. Observations of local flamingo bird populations and their breeding activity, however, are leading biologists towards a conclusion that pumping brine from salars is affecting the local water tables, climate, vegetation, and wildlife.

Since 2000, local community groups along with environmental activist groups have been raising their voices over concerns that brine extraction is affecting the ecology of the surrounding areas. Their voices, however, have not been listened to. The push for battery-powered vehicles has just been too intense. The geographic area around the salars is hydrologically connected and balanced. It is one big ecosystem. For every liter of brine that is pumped from a salar, the underground soil will recharge that volume from sub-surface moisture emanating from the nearby mountain ranges. This ultimately means that brine extraction is upsetting the local water tables. Farmers situated near the mountain ranges are reporting decreased amounts of fresh water in their water wells.

To counter the arguments from environmental groups, some companies, attempting to establish themselves in the Lithium Triangle, are now proposing to avoid the evaporation process by using an electrochemical extraction process. At first glance, this sort of proposal sounds clean and environmentally friendly. However, the brine must still be extracted from the salar and the local water tables will still be disrupted. Moreover, any electrochemical extraction process will require significant amounts of electricity from coal or natural gas generating plants to power the electrochemical refining process.

Electrochemical extraction process details are not openly discussed by companies promising to use them. Judging by the number of scientific papers online that are mentioning electrochemical lithium extraction, it appears that this refining approach is going to be a hotly contested issue in the coming years. Investors considering buying shares in lithium extraction companies should not regard the electrolytic refining of lithium hydroxide to be a generic term, nor should they assume that electrochemical refining is straightforward. Process variables including operating temperature, current, voltage, anode design, and cathode design all will have to be taken into consideration.

Meanwhile, extraction of lithium from salars in the Lithium Triangle will continue. The promotion of electric vehicles will continue. The complaints from local indigenous groups and environmental activist groups will continue. **The environmental damage in the Lithium Triangle will continue as the world sleepwalks towards the utopian goal of electric cars in every driveway.**

Lithium From Abandoned Oil Wells
Lithium mineralization is also found in minor concentrations in deep oil wells that have intersected underground aquifer formations. The water in the underground aquifer has over time leached lithium ions from the surrounding porous rock formations. There are numerous companies in the U.S. and Canada attempting to monetize the extraction of lithium from these former oil wells by using ion exchange (electrochemical) processes. As of mid-2023, none of these companies has achieved commercial success. It is one thing to draw a fluid sample from an abandoned oil well, subject it to laboratory extraction, craft an exciting press release to promote the share price and another matter to scale up to a large economic scenario. The lithium content in a typical abandoned deep well is around 250 ppm. Compare this to 1,400 ppm in a South American salar situation, 2,000 ppm in a clay deposit in Nevada, and as much as 13,000 ppm in a hard rock deposit and the immediate conclusion is that capturing lithium mineralization from abandoned oil wells is not economically feasible. Add in the fact that, even though abandoned, the wells in the oilfield are still owned by an oil company. This means a lithium project operator will have to collaborate with the oil company through a revenue sharing agreement to gain access to the wells. This further could erode the economics of

lithium extraction. Brine projects are not as simple as just bringing fluids up from depth.

An example of a brine lithium project is that of Canadian-based Standard Lithium (TSXv:SLI) and its Arkansas deep well project. Standard Lithium will make annual payments to the company owning the sub-surface oil and gas rights. The aquifer formation containing the lithium brine is 8,800 feet below surface. Pumping up brine in sufficient volume to feed a processing plant will entail drilling 23 deep extraction wells, 24 deep waste-disposal wells, and laying over 20 miles of pipeline. The estimated capital cost of this project was $870 million in the 2021 economic assessment filed with Canadian securities regulators. The extraction process will rely on the *DLE process*—a nanofiltration method. The DLE process is reported to be able to generate a lithium chloride concentrate containing 60,000 ppm lithium chloride. A late 2023 press release from Standard Lithium claims that 58 million liters of brine has been DLE processed to produce a filtrate containing only 10,000 ppm lithium chloride. A share price chart tells the rest of the story. In early 2023, share price was around $5. In early 2024, share price is languishing at the $1.50 level. The Q1 2024 financials showed a cash burn of over $20 million. In mid-2023, the company engaged French banker PNB Paribas to secure project financing. As of early 2024, no progress had been made on the financing front.

Peak Lithium?

A 2012 paper by Australian researchers Mohr and colleagues summarized the lithium resource estimates cited in published academic literature. The figures in previous literature varied from 38 million tonnes to 50 million tonnes. In a mining extraction scenario, not all of a resource quantity can be economically extracted. Mohr and his colleagues estimate that 23.6 million tonnes of lithium (carbonate or hydroxide format) will ultimately be recoverable around the globe. The United States Geological Survey (USGS) estimates the global reserves of lithium to be 22 million tons (24 million tonnes). This estimate aligns to the work of Mohr and colleagues.

In a previous chapter, it was estimated that a 4680-battery with 960 cells will contain 32 kgs of lithium. Between 2023 and 2030, forecasts shown on the *www.evvolumes.com* website suggest that nearly 186 million electric passenger vehicles will be made by 2030 worldwide. Estimates show that 35 million hybrid vehicles will be made. Assume that 40% of these non-hybrid vehicles will use batteries of the 4680 design each containing 32 kgs of lithium. Assume the other 60% will use the LFP design of battery with 50 kgs of lithium. Assume the hybrid vehicles will use battery designs containing 8 kgs of lithium. Making 186 million electric vehicles and 35 million hybrid vehicles between 2023 and 2030 will use nearly 8.25 million tonnes of lithium. **This means that in about 20 years (by 2043), the globe will have largely depleted its lithium reserves as the political dream of electric vehicles unfolds.**

The World Economic Forum (WEF) takes a more aggressive stance. The WEF makes an argument that in order to achieve a net-zero environment, the globe will need two billion EVs on the roadways by 2050. The calculations just presented point to peak scenarios for battery metals. The idea of two billion electric vehicles on the roadways by 2050 is not even plausible. In fact, the idea is laughable. This shows how idealistic and out of touch the highly influential WEF really is. Little, if any, mathematical calculation has gone into their forecast.

What is disturbing is how our political leaders and academic minds have chosen to blindly accept the WEF logic while the storm clouds of peak metal scenarios build on the horizon.

SILVER

In numerous locations around the globe, magma at depth has absorbed silver mineralization along with sulfur mineralization from the surrounding country rock. Magma coming to near the surface has given rise to deposits containing silver sulfide mineralization. A prolific area for silver mineralization is Mexico, where tectonic plate subduction has formed the Sierra Madre mountains. Magma coming to near surface was enriched in silver mineralization at levels of several hundreds of ppm.

When one thinks of silver, thoughts of silver coins come to mind. However, silver is seldom used in coins anymore. It is an industrial metal. As society adopts more technology, silver will become a highly sought after commodity.

The average cell phone contains 0.2 grams of silver. Multiply that by the estimated 16 billion cell phones existing on the planet and the result is 3.2 million kgs of silver. Global cell phone makers are producing around 1.7 billion new phones a year and consuming around 340,000 kgs of silver in the process.

The average computer contains one gram of silver. Multiply that by the estimated two billion computers on the planet and the result is two million kgs of silver. Global computer makers are creating an estimated 300 million new units per year and consuming 300,000 kgs of silver in the process.

The average 200-Watt solar panel (3 feet x 5 feet) that you see perched on residential roof tops contains 20 grams of silver. In 2022 alone, the world added 130 billion Watts (130 GW) of additional solar capacity (equivalent to 650 million 3 foot x 5 foot panels). This equates to 13 million kgs of silver.

Silver is used in electric vehicles as a coating on electrical connectivity plug ends. It is estimated that a typical electric vehicle uses between 25 and 50 grams of silver. Continuing on with the 186 million vehicle sales projection figure, the electric vehicle industry between 2023 and 2023 stands to consume between 4.5 and 9 million kgs of silver.

In a typical year, approximately one million kgs of silver is brought to market by miners in the U.S. There is a reported one million kgs of silver in the COMEX warehouse (owned by the Chicago Mercantile Exchange). There is a reported 2.2 million kgs of silver inventory at the Shanghai Metal Exchange. These numbers suggest that silver, the industrial metal long regarded as rather boring by investors, is fast approaching the point of scarcity on its way to becoming a precious metal.

There will not be enough silver to satisfy all of the technological end uses (including electric vehicles) that the world envisions.

RARE EARTH METALS

An estimated 150 to 200 million years ago, there was a tectonic plate subduction event in an area of the Asian continent that today forms part of China. The subduction created a magma melt at depth which became enriched in lanthanide elements such as lanthanum, cerium, neodymium, and praseodymium. In addition, the magma melt became enriched in phosphate, silicate, and carbonate mineralization. Heat and pressure forced the magma to near surface in a granite rock intrusive formation. Over the following millions of years, the granitic rocks along with the lanthanide minerals were weathered away. The weathered lanthanide minerals accumulated in clay formations. China controls over 90% of the world's lanthanide type minerals. Given the scarcity of these elements, they are termed *rare earth elements.*

To better appreciate rare earth elements, consider the practical example of an electric motor. Inside the motor are stationary electric coils of copper wire positioned around a moveable magnet (rotor). As electrical power is supplied to the electric coils, the principle of electromagnetic induction causes the moveable magnet assembly to rotate. If this assembly is attached to a shaft, the rotating shaft can be used to provide mechanical energy to an attached device. The magnets in the rotor are typically made from iron alloyed with nickel, zinc, or manganese. Electric motors can be made lighter in weight by replacing the iron alloy magnets with magnets made from alloys containing rare earth elements.

While weight may not be a concern for the electric motor in a device such as a washing machine, weight is a concern for automotive designers who incorporate various small electric motors into vehicles to power windows up and down, to move seats ahead and back, and to turn the windshield wipers on and off. Reducing the overall weight of a vehicle will translate into improved fuel efficiency.

Vehicle weight is a primary concern for electric vehicles. Electric vehicle consumers expect a decent driving range between battery charges. The weight of the overall vehicle is critical in achieving longer driving distances. To reduce the weight of the vehicle, motors based on rare earth magnets are quite often used. Rare earth magnets are about 1.5 times as heavy as ferrite magnets, but rare earth magnets have a magnetic strength that is up to five times stronger than ferrite magnets. Therefore, smaller, stronger magnets can be used to reduce vehicle weight. On the Periodic Table of the Elements, the rare earth material of choice for most magnets is element number 60—neodymium. To create a magnet, the neodymium is alloyed with iron and boron.

Since the 1980s, China has annually produced over 90% of the world's supply of rare earth minerals. As discussed in the earlier chapter on metal hydride batteries, China demonstrated its power over the rare earth metals market in 2010 when a political spat with Japan resulted in China threatening to retaliate by curbing exports of rare earth metals to Japan. This geopolitical example has not been forgotten. Countries around the world are now encouraging the mining industry to identify new domestic sources of rare earth mineralization.

Continuing with the projection of 186 million electric vehicles and 35 million hybrid vehicles to be made by 2023, at a usage of 3 kgs per vehicle, the mining industry will have to provide over 600,000 tonnes of neodymium. This eclipses the 2023-2030 projected figure of 70,000 tonnes of total neodymium consumption for all other industrial applications. China will no doubt use this scenario to its geopolitical advantage.

- If electric vehicle sales amount to the projected 186 million units by 2030, the planet will become exhausted of its cobalt and nickel supplies. The decline of copper and manganese reserves will come under duress.
- The South American lithium salar deposits have potential negative environmental ramifications as water tables are affected. The continued push towards vehicle electrification will see a rapid decline in global lithium resources.

- The continued push towards vehicle electrification will see silver come into short supply.
- China controls the global rare earth market. The rare earth metal neodymium is a key component in electric motors that empower the drive train on an electric vehicle.
- Continued promotion of electric vehicles will hasten peak battery metal scenarios and increased geopolitical tensions.

THERE IS NO FREE LUNCH

Electric vehicle proponents will argue that battery powered cars produce no tailpipe exhaust emissions; therefore, electric cars are saving the planet. But electric vehicle proponents rarely consider where the energy comes from to create the electricity that allows a driver to plug in his/her car at a charging station. It takes energy to make electrical energy.

Whether it is heating crushed ore in a kiln at 1100°C, heating lithium-bearing brine material to 90°C, or using 35 kW·hrs to recover one kg of lithium by electrolysis of a solution containing lithium mineralization—energy to make the battery components has to come from somewhere. Once the lithium battery to power an electric vehicle is installed, additional electrical energy to charge that battery then has to be created.

The energy that we take for granted in our society is not free. There is no free lunch.

Our world revolves around the principles of *thermodynamics* (the science of heat, work, temperature, and energy). Unfortunately, the math and science of thermodynamics is complicated to understand and therefore consumers can be easily persuaded into thinking that the purchase of an electric vehicle will save the environment from the ravages of climate change.

FIRST LAW OF THERMODYNAMICS

The First Law of Thermodynamics states that energy cannot be created or destroyed. It can only be converted from one form to another.

Our world, including our solar system, is a closed system; no external influences add energy to our world. The energy within our closed system can only be transferred within the system in two ways: heat and work.

The mathematical statement of the First Law of thermodynamics is: $\Delta U = q + w$.

In other words, the change of energy (ΔU) of a closed system equals heat energy (q) plus work energy (w). The change in energy of the closed system, can take the form of kinetic energy, potential energy, or thermal energy.

In the context of electric vehicles and batteries, consider the burning of thermal coal at a generating station. The coal is burned and the chemical energy contained in the coal is turned into thermal energy which boils water and creates steam. The energy in the steam is used to turn a turbine generator which makes electricity. The electricity is used to charge a battery that empowers the electric vehicle.

SECOND LAW OF THERMODYNAMICS

The Second Law of Thermodynamics hinges around the concept of *entropy* (the randomness of a closed system). The Second Law states that the entropy of a closed system can never decrease over time; entropy can only tend towards equilibrium of the system.

Consider the example of a closed system that consists of a rubber ball sitting on the edge of a table. The ball rolls off the table and begins to bounce. As the ball hits the floor, the ball material momentarily distorts, and the impact creates heat frictional energy. The energy remaining in the ball causes it to rebound, but not to the same height as it had been initially. The quality of the energy remining in the ball after each bounce diminishes and the ball will eventually stop bouncing as equilibrium is reached.

As another example, consider a closed system comprised of a hot object and a cold object. As the two objects touch, heat energy flows from the hot object to the cold object. The temperature of the hot object decreases and the temperature of the cold object increases. The result will be an equilibrium situation where the temperatures of the two objects are equal. The Second Law of thermodynamics cannot be beaten; energy is dissipated in all naturally occurring processes. All of these processes have an efficiency limit.

In the context of electric vehicles powered by lithium batteries, energy is used to power mining equipment and processing plants that remove minerals from beneath the ground. More energy is then used to refine these minerals into metals that are used to construct the batteries. Energy is used to create electricity to charge the batteries. All of the activity surrounding mining, refining, battery making, and electrical generation has an efficiency limit. Heat energy will be liberated in all of these activities. These activities inefficiently and irreversibly increase the entropy of our closed solar system. **As more entropy is created, our planet moves towards an equilibrium state where human life can no longer be sustained. To slow the rate of entropy creation, the eight billion of us on the planet need to reduce the amount of energy and materials being consumed. Isn't this what researchers like Jurgen Randers have been saying since the early 1970s?**

- The First Law of Thermodynamics states that energy cannot be created or destroyed. It is transferred from one form to another.
- The Second Law states that the entropy of a closed system can never decrease over time; entropy can only tend towards equilibrium of the system.
- Taken together, these two Laws tell us that the energy that we take for granted in our society is not free.
- Electric vehicles will not save the planet. There is no free lunch.

ELECTRICITY FROM THE WIND?

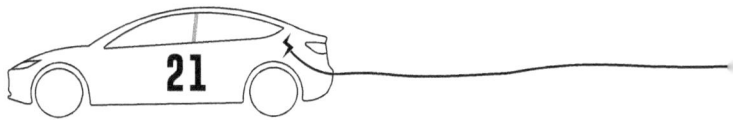

Green energy proponents will argue that the wind is a source of free energy to create electricity from. All we need to do is build wind turbines, capture this wind energy, convert it into electricity, and the world will be a wonderful place. Unfortunately, the First Law of Thermodynamics gets in the way of this idyllic thought process.

Scan the horizon near where you live and you might spot a wind turbine off in the distance. The rotating blades of the wind turbine are a classic example of the First Law of Thermodynamics at work. The energy of the blowing wind causes the blades of the turbine to turn. The rotating blades turn a gearbox which turns a small generator to create electricity which is fed into the electrical grid. In the expression for the First Law of Thermodynamics ($\Delta U = q + w$), ΔU is the wind energy captured by the blades, q is the energy wasted due to friction of the parts in the rotating gearbox, and w is the work energy available to turn the generator to create electrical energy for the grid.

Wind turbines do not materialize magically. They have to be manufactured somewhere. Here again one can see thermodynamics in play. It takes significant amounts of thermal energy (ΔU) to melt iron ore and coal to create the pig iron that is used to create the 50 tonnes of structural steel needed for making the sub-surface base structure to support a wind turbine. The vertical column that supports the turbine will be made from nearly 12 tonnes of steel. Making all this steel generates about 90 tonnes of carbon dioxide.

The base of a wind turbine requires nearly 500 tonnes of concrete. Making the cement for this amount of concrete by mining, crushing, grinding, and calcining limestone takes significant amounts of thermal energy. Making the estimated 70 tonnes of cement for 500 tonnes of concrete will release over 60 tonnes of carbon dioxide into the atmosphere.

It takes significant amounts of thermal energy to melt silica sand to make the fiberglass used in the fabrication of turbine blades. Energy is further used to transport the various wind turbine components from the point of manufacture to the site where the turbine will be erected. Energy is further used in mining and fabricating the rare earth metals for the permanent magnets that comprise the generator unit atop the turbine. In order for a wind turbine to recover the equivalent energy required to create and install it will take several years.

Once operational, a rotating wind turbine is far from efficient in that the rotating blades do not capture all the available wind kinetic energy (ΔU). In 1919, German physicist Albert Betz concluded that at most only 59% of the kinetic energy from wind will be captured by a spinning turbine blade. This observation is now called *Betz's Law*. The U.S. Environmental Protection Agency (EPA) goes beyond the Betz efficiency limit estimating that a typical wind turbine might only be 40% efficient at generating electricity.

If a turbine only removes a portion of the wind energy, a more nuanced issue that researchers are now investigating is whether the reduced-energy wind flow leaving the turbine has a negative effect on the microclimate in the area immediately surrounding the turbine.

There is no free lunch. Creating electrical energy from the wind is not efficient. Thinking that if we dot the landscape with wind turbines, we can easily create cheap electrical energy to charge the batteries in electric vehicles is flawed logic.

- Both the First Law of Thermodynamics and Betz's Law remind us that wind turbines are not efficient devices.

CHARGED!

ELECTRIC VERSUS INTERNAL COMBUSTION

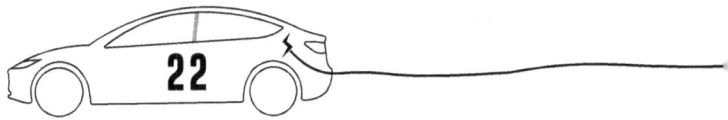

22

At a coal-fired generating station, coal is fed into a combustion furnace where the heat of combustion causes the temperature of a water-filled pressure boiler to rise. Eventually the water in the boiler will reach the boiling point and steam will be generated as the water changes from liquid to vapor phase. The energy to raise one kilogram of water from room temperature to the boiling point is 335 kiloJoules (kJ). To turn that hot water into steam will require a further 2,260 kJ of energy. The high-pressure steam released from the boiler in a controlled manner is directed towards a turbine. The rotating turbine is connected to a generator that creates electricity. The electricity is fed into transmission lines which carry it to nearby towns and cities. In these towns and cities, some of this electricity might be directed to a charging station where electric vehicle owners can charge their vehicle battery. In the context of the First Law of Thermodynamics, the change of energy (ΔU) in the coal-fired generating plant system is expressed by q (the heat of combustion of the coal), plus w (the work energy that rotates the generator shaft and creates electricity).

A typical coal fired generating plant is only around 40% efficient. Some of the combustion energy from the burning coal is consumed to bring about the phase change of the water from liquid to steam vapor. Some of the combustion energy is lost in the form of heat radiating away from the combustion furnace and heat radiating away from the steam turbine unit. In addition, some heat is lost up the flue smokestack. A small portion of the coal does not properly combust and is collected in the form of waste fly ash from the flue stack.

The electricity from the generating plant then must pass through a step-up transformer to raise its voltage. This entails a heat loss of 1-2% of the electric energy that has just been created. The high voltage electricity is then moved through transmission lines to nearby population centers. Heat losses are incurred along the entire length of the transmission lines. This wastes between 2-4% of the electric energy that has been generated. The electricity from the high-voltage transmission lines is then passed through step-down transformers to reduce the voltage to a level needed by the end user. This entails a heat loss of 1-2%. Add these losses up and a typical power transmission line is at best 95% efficient.

A NUMERICAL EXAMPLE TO MAKE SENSE OF IT ALL

Electric vehicle proponents like to cite the 85% efficiency of a typical electric car. Consider the Tesla Model 3 with an EPA estimated driving range of 536 kms (333 miles). At highway speeds of 100-110 km/hr, this range decreases to around 400 kms (250 miles) according to various online blogs where Tesla owners have complained about driving range. A fully-charged battery pack in a Tesla Model S will contain around 82 kW·hrs. This equates to 4.87 kms of driving range per kW·hr of energy.

As noted in the definitions at the beginning of the book, one Joule is the work required to produce one Watt of power for one second. The 82 kW·hrs in the charged battery equates to 295 Megajoules (MJ).

The U.S. Energy Information Administration (EIA) reports that the average thermal energy contained in U.S. coal is 45 kJ per kg of coal. In a generating plant that is 40% efficient, one kilogram of U.S. thermal coal will generate 18 kJ of energy. This quantity of energy equates to 0.005 kW·hrs. To generate 82 kW·hrs to fully charge the battery pack, 16,400 kgs of coal will need to be combusted at the power plant. Assume that the efficiency of transmitting the electrical energy through the grid system is 95%. To compensate for this efficiency level, the amount of coal to be burned now becomes 17,000 kgs. Factor in that a typical vehicle charging station is perhaps 90% efficient, and the amount of coal to be burned becomes closer to 19,000 kgs. Expressed a different way, 19,000 kgs of coal must be burned to create 360 MJ of energy to ultimately allow the electric vehicle owner to charge the battery with 295 MJ of energy so he or she can drive the vehicle car 400 kms along the roadway.

The CO_2 created in burning the coal to make enough electricity to charge the battery on a typical electric vehicle is around 5,800 kgs.

Let's imagine this same vehicle was powered by an internal combustion gasoline engine. One barrel of crude oil (42 U.S. gallons) contains on average 1,665 kW·hrs of energy (6,000 MJ). The crude oil refining process has seen major efficiency improvements over the past couple decades and only wastes about 7% of the energy content in a barrel of oil. A barrel of oil passing through the refinery will produce almost 20 gallons of automobile gasoline with each gallon having an energy content of 35 kW·hrs (127 MJ).

A typical oil refinery will generate 12 grams of carbon dioxide per MJ (0.277 kW·hrs) of energy produced. The creation of 20 gallons of gasoline from a single barrel of crude oil will produce 1,524 grams (1.524 kgs) of carbon dioxide.

Let's tie the parts of this argument together. The average gasoline powered vehicle is about 35% efficient. Burning one gallon of gasoline at this efficiency level will deliver about 44 MJ of energy which is equal to about 12 kW·hrs. The 360 MJ of coal energy to move the Tesla vehicle in our example 250 miles (400 kms) equates to about eight gallons (30 liters) of gasoline to deliver an equivalent amount of energy. This equates to 31.25 miles per gallon of fuel burned. As a point of reference, two long- distance trips in 2023 with my wife's 2009 BWM 128i both delivered 38 miles per gallon in fuel consumption.

The conclusion—a Tesla model 3 electric car is no more or no less efficient than an older model BMW. Electric vehicles are not the planet-saving free lunch that people think they are.

In terms of emissions, a typical inefficient coal fired generation plant will produce 150 grams of carbon dioxide to generate one MJ of energy. The average gasoline powered automobile will generate about 73 grams of carbon dioxide per MJ of energy created. Factor in the carbon dioxide created by the crude oil refining process, and the coal fired generating plant still produces more carbon dioxide.

The overall cost of empowering an electric vehicle along the roadway might appear cheaper if electricity is priced at 20 cents per kW·hour but both gasoline and coal are hydrocarbons. Electric vehicle advocates should move beyond the excitement of plugging

a vehicle into a charging station and then boasting about how they are saving the planet and saving money.

- When one factors in the thermodynamic efficiencies of coal generating plants, electrical transmission lines, and charging stations, and compares the amount of energy in terms of liters of fuel equivalent, it becomes apparent that an electric car such as a Tesla model 3 is no more or no less efficient than an older model BMW 128i. There is no free lunch.

CHARGED!

VEHICLES POWERED BY HYDROGEN FUEL CELLS?

23

If a continued push towards electric vehicles creates a depletion of mineral resources, is there another way of providing vehicle mobility that would meet with the political agenda of zero-emissions? The short answer is—yes, hydrogen fuel cells.

Using hydrogen as a fuel to propel vehicles and other devices traces its origins back to efforts at General Electric in 1960 when the company developed a small fuel cell to potentially power ships for the U.S. Navy. This small fuel cell design would also be used to power several of the lunar modules for the Gemini space program.

Efforts to create a fuel cell to power a vehicle proved more challenging. Prototypes developed by American and European scientists in the late 1960s and early 1970s were found to be lacking in terms of current densities, voltages, and continuous operating time between refueling. What did emerge from this research, however, was the conclusion that hydrogen is a desirable fuel for fuel cells given its fast reaction time and that its process by-product is environmentally-benign water.

A fuel cell is comprised of an anode, a cathode, a catalyst, and a porous membrane. Hydrogen fuel at the anode side of the cell releases electrons (e^-) and protons (H^+ ions). The chemical reaction is:

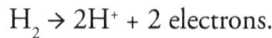

$$H_2 \rightarrow 2H^+ + 2 \text{ electrons.}$$

The protons (H^+ ions) travel through the porous membrane where they interact with an oxidizing substance (oxygen from the air) at the cathode side of the cell. The chemical reaction is:

$$\frac{1}{2} O_2 + 2H^+ + 2 \text{ electrons} \rightarrow H_2O.$$

Meanwhile, the electrons travel through the fuel cell circuitry to the cathode side of the device where they too join the chemical reaction with the protons. This movement of electrons through the fuel cell circuitry creates electrical energy which can be used to power to an electric motor that turns the vehicle drivetrain.

The porous membrane in fuel cells is often a material called Nafion 1170 which was developed by chemical company, Chemours. The catalyst material in fuel cells forms part of the anode structure and is typically an alloy of platinum metal. The catalyst material ensures that the hydrogen fuel quickly and efficiently breaks into its electron and proton constituents.

As environmentally friendly as fuel cells may sound, as of early 2024, battery power continues to steal the limelight on the stage. There are only three models of fuel cell vehicles available: the Honda CRV, the Toyota Mirai, and the Hyundai Nexo. Each of these models generates around 180 horsepower, and will accelerate from 0 to 60 miles per hour in between 7 and 8 seconds. This type of performance makes them ideal for city commuting. However, as noted in an earlier chapter, North American consumers have a fixation on performance. A driver seeking rapid acceleration can go from zero to 60 miles per hour in around 3.7 seconds in a Chevrolet Camaro that cranks out nearly 650 horsepower. A Tesla Model Y will go from zero to 60 miles per hour in just over five seconds.

Another barrier to further fuel-cell models being made available is the lack of hydrogen filling station infrastructure. The U.S. Department of Energy (DOE) website states that as of early 2024 there are only 57 hydrogen filling stations in the U.S.–all in California. If passenger vehicles powered by fuel cells are to become mainstream, it is imperative that work begin to develop hydrogen infrastructure. Facilities will need to be created to generate the hydrogen. Retail gasoline stations will have to be equipped with hydrogen dispensing facilities and a delivery system will have to be created to get the hydrogen to these individual stations.

There are multiple designs of fuel cell. The design appropriate for powering vehicles is the *polymer electrolyte membrane* (PEM) design. This design is also referred to as the *proton exchange membrane fuel cell.* The PEM fuel cell gives high fuel consumption efficiency, near zero noise pollution, and an exhaust by-product of nothing more than water.

The efficiency of a PEM fuel cell is 83%. This is calculated by way of the First Law of Thermodynamics which states: $\Delta U = q + w$. The overall combined reaction equation for a fuel cell is: $H_2 + \frac{1}{2} O_2 \rightarrow H_2O$.

The change in energy (ΔU) for this overall reaction is 285.8 Joules. The energy available for work (w) is 237.1 Joules. The efficiency of a

PEM hydrogen fuel cell is thus 237.1/285.8 = 82.9%. This figure is certainly better than a gasoline-powered internal combustion engine that operates at around 40% efficiency.

Hydrogen is not a fuel in the same sense that crude oil is. Hydrogen must be obtained through chemical and thermal reactions with materials that contain hydrogen atoms. Obtaining hydrogen can be accomplished in one of three ways: steam reforming of natural gas, partial oxidation of natural gas, or electrolysis of water.

Due to their low operating temperature (around 70°C), PEM fuel cells cannot directly use hydrocarbon fuels, such as natural gas, liquefied natural gas, or ethanol.

STEAM REFORMING

In the steam-reforming process, natural gas reacts with steam under high pressure in the presence of a catalyst. The reaction produces hydrogen, and carbon monoxide as follows:

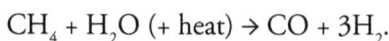

$$CH_4 + H_2O \text{ (+ heat)} \rightarrow CO + 3H_2.$$

The carbon monoxide is then exposed to additional steam. The reaction produces carbon dioxide and hydrogen:

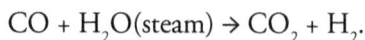

$$CO + H_2O(steam) \rightarrow CO_2 + H_2.$$

The combination of the two reactions can be expressed as:

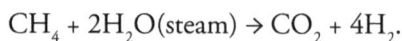

$$CH_4 + 2H_2O(steam) \rightarrow CO_2 + 4H_2.$$

The important factor to note is that heat energy must be input to make these reactions feasible. It takes energy to make energy. The energy input to make this overall reaction proceed is 165 kJ per mol of natural gas. The CO_2 in the above reaction equation is vented to the atmosphere and the hydrogen is captured. The practical efficiency of the steam reforming process is around 70%. The combined efficiency of the steam reforming process and subsequent fuel cell would therefore be around 58%. There is no free lunch as the first two Laws of Thermodynamics remind us. Creating hydrogen fuel by steam reforming takes energy and adds to the greenhouse gas emissions already being generated on the planet.

PARTIAL OXIDATION

In partial oxidation, natural gas reacts with a small amount of oxygen to produce carbon monoxide and hydrogen:

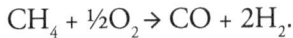

$$CH_4 + \tfrac{1}{2}O_2 \rightarrow CO + 2H_2.$$

The carbon monoxide is then exposed to additional steam. The reaction produces carbon dioxide and some hydrogen:

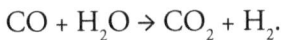

$$CO + H_2O \rightarrow CO_2 + H_2.$$

The partial oxidation process does not require an input of energy. The overall reaction actually generates heat. The efficiency of the partial oxidation process is about 80%. The combined efficiency of the steam-reforming process and subsequent fuel cell would be about 66%. As with steam-reforming, the partial oxidation process generates carbon dioxide, however the efficiency of this process is greater than the steam reforming approach.

ELECTROLYSIS OF WATER

Hydrogen fuel can also be created by adding enough energy to a molecule of water (H_2O) to break the molecule into its constituent parts. The reaction steps are:

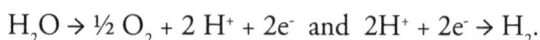

$$H_2O \rightarrow \tfrac{1}{2} O_2 + 2\,H^+ + 2e^- \text{ and } 2H^+ + 2e^- \rightarrow H_2.$$

The electrolysis of water is 78% efficient. The energy required to dissociate one mole of water is 237 kJ. The energy contained in one mol of hydrogen gas is 302 kJ. The efficiency of the hydrolysis process can be calculated as 237/302 = 78%.

The use of hydrogen fuel cells to power cars raises the issue of the platinum catalyst material. The World Platinum Investment Council (WPIC) raised eyebrows in early 2023 when it announced global platinum demand will outstrip supply in 2023 by 983,000 ounces (27,868 kg) largely due to ongoing electricity shortages in South Africa, which produces and refines 70% of the world's platinum. Concerns over platinum availability bring into question the viability of fuel cells to empower vehicles on a wide scale.

And then there is the issue of overall efficiency. Even though the chemical efficiency of making hydrogen is 78% efficient, getting the hydrogen to individual retail filling stations by way of tanker truck is not efficient. According to Norwegian-based DN Media Group's website *www.hydrogeninsight.com*, efficiency figures from both Europe and the U.S. Department of Energy suggest that creating hydrogen by electrolysis, then liquefying, transporting, storing, and distributing the hydrogen fuel to car drivers to use in fuel-cell vehicles would be 23% efficient. There is no free lunch. It takes energy to make, transport, and store energy. As clean energy thought leader Michael Liebreich of

Liebreich Associates told attendees at the World Hydrogen Congress in Rotterdam in June 2023, "the view that hydrogen is a silver bullet or a Swiss Army knife capable of decarbonizing everything from heating to transport to heavy industry and power generation is dangerous."

Despite the low overall efficiency of hydrogen fuel cells, the California Air Resources Board (CARB) plans to publish a hydrogen fuel evaluation by June 1, 2024. The evaluation will model how hydrogen supports the decarbonization of the transportation sector. The evaluation will focus on the number of hydrogen filling stations needed to support varying amounts of hydrogen fueled vehicles on the roads, and the operating standards of those stations. It will be interesting to see if California does press further ahead with hydrogen as a vehicle fuel and if the automakers can respond with hydrogen powered vehicle models. As of early 2023, some automakers claimed to be working on fuel cell designs.

- Vehicles empowered by hydrogen fuel cells that produce an exhaust emission of nothing more than water sound clean and environmentally friendly. There are three vehicle models currently available to consumers that run on hydrogen fuel cells.
- The efficiencies of creating hydrogen and distributing it to retail filling stations are chemically and thermodynamically challenging. It takes energy to make energy.
- Even if our politicians decided to embrace hydrogen, it would take many years to develop a hydrogen infrastructure in North America and Europe.

CHARGED!

NATURAL GAS TO POWER VEHICLES?

24

Is natural gas as a fuel to run a power generating plant a more efficient option than coal? There certainly is an abundance of natural gas available. The U.S. Energy Information Administration (EIA) states that there are proven natural gas reserves of more than 600 trillion cubic feet—equal to 18 years of consumption at today's usage rates. There is an even larger amount of what is deemed unproven resources—2500 trillion cubic feet. Assuming that 80% of this unproven figure could ultimately be extracted, that would be enough natural gas to supply the U.S. economy for another 62 years.

A natural-gas-fired power plant comprises three main parts: a compressor, a combustion chamber, and a turbine/generator. The compressor does what the name implies—it compresses a mixture of air and natural gas and feeds the mixture to the combustion chamber. The air-fuel mixture is combusted at near 2000°C producing a high temperature, high pressure stream of exhaust gas that then turns the blades on a turbine. In turn, the rotating turbine blades provide kinetic energy which drives the compressor to both draw more pressurized air-fuel into the combustion chamber, and spin the generator to produce electricity.

There are two general designs of gas-fired generating plant: *simple-cycle*, and *combined-cycle*. A simple-cycle gas turbine operates at an efficiency of around 30%. The inefficiency is due to the combusted exhaust that is not consumed in turning the turbine being emitted to the atmosphere and wasted. A combined-cycle plant will capture the unused exhaust gas and use its thermal energy to heat water to make steam to turn a turbine. The Second Law of Thermodynamics comes into focus here, with the overall efficiency of a combined cycle plant being near 60%.

A typical combined cycle natural gas plant will have a capacity of around 400 MW of power output. Assuming the plant runs at this maximum output level every day for 12 months with no shutdowns, the electricity generated will amount to 3.5 million MW·hrs. This in turn will produce near 1.3 million tonnes of carbon dioxide emission. A typical coal-fired generating plant producing this same 400 MW of power will emit around 2 million tonnes of carbon dioxide.

So, are combined cycle natural gas-fired generating plants the answer to more efficient power generation? They are certainly more efficient than coal-fired plants. They are more efficient than wind turbines. And they do not have the negative baggage of nuclear waste. However, the issue of depleting the planet's resources of strategic metals in making batteries still lingers. **Depleting natural gas reserves to make electrical power so that society can be driving electric vehicles by 2035 also lingers as an issue. Natural gas is a hydrocarbon and burning it creates carbon dioxide. There is no free lunch. Jurgen Randers and the IPAT Model warned us of this years ago.**

- Combined cycle natural gas fired generating plants have greater efficiency than coal fired plants.
- EIA estimates indicate the U.S. has nearly 18 years of natural gas reserves remaining, based on current consumption levels. However, when these natural gas reserves are gone, they are gone for good. The EIA estimates there are 62 years of natural gas resources remaining, but this figures is unproven.

CHARGED!

ELECTRICITY FROM NUCLEAR POWER?

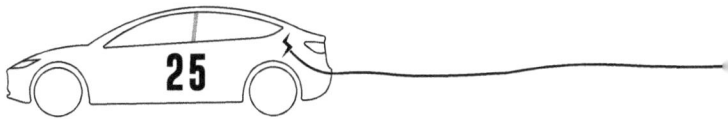

25

Set aside the efficiency concerns of hydrocarbon-fueled electricity generating plants for a moment. Set aside concerns over battery metals for a moment. What if electricity could be generated from a source that did not contribute to greenhouse gas emissions? What if this electricity could be used to charge electric vehicle batteries?

Elected officials in four Canadian provinces are now talking in earnest about using nuclear energy to generate electricity. In the past, even the mention of building nuclear power stations has been the downfall of more than one politician as angry electorates have stomped their feet in disgust and invoked the NIMBY (not in my back yard) argument. But what if there was an aggressive focus on generating environmentally-friendly electricity from nuclear power? Could voters be swayed?

A nuclear plant is similar to a coal-fired generating plant in that the energy extracted from a fuel source is used to heat water until it turns into steam. The high-pressure steam is used to turn a turbine. The rotation of the turbine drives a generator that produces electrical energy.

The general inefficiency that applies to a coal-fired generating plant also applies to a nuclear plant. The energy to raise the temperature of one kilogram of water from room temperature to the boiling point is 335 kJ. To turn that hot water into steam will require a further 2,260 kJ of energy to bring about the phase change from liquid to vapor. The excess steam not used to drive the turbine is cooled back to liquid phase in a heat exchanger and directed back to the boiler where the heating process is repeated. The average nuclear facility thus has an efficiency of around 40%, not dissimilar to a coal-fired facility. The only difference is the nuclear plant does not produce greenhouse gas emissions.

There are two types of reactor design: slow and fast. The nuclear fuel contained in the fuel rods at a slow nuclear facility comprises pellets of enriched U-235 material. The nucleus of an isotope of U-235 contains 92 protons and 143 neutrons (92 + 143 = 235). The U-235 material originates in mined uranium oxide material (U_3O_8) which contains about 0.7% by weight of U-235 and 99.3% of U-238 isotopes. To make U_3O_8 (yellowcake) material useful as a fuel source for a reactor, it is first converted to uranium hexafluoride (UF_6) in a gaseous state. The gaseous UF_6 material undergoes centrifuge enrichment to eliminate the U-238 content and concentrate the U-235 content. The U-235 material is then cooled back to solid phase, ground into a powder, converted into oxide format (UO_2) and pressed into pellets. The pellets are packed into fuel rods which are inserted into the reactor chamber.

The isotopes of U-235 are unstable. If an isotope of U-235 is hit by another isotope, it will absorb a neutron from the colliding isotope. The energy absorbed from this collision induces the U-235 isotope to split into two parts. The energy released from this split is called *fission energy.* The fission reaction, in turn, releases a neutron which in turn hits another U-235 isotope, causing it to split and release energy. This is why the analogy of billiard balls on a table all banging into each other is often used to describe the fission process. To ensure that the neutrons released end up impacting other U-235 isotopes at a proper speed to bring about fission, the neutron velocities must be slowed.

This is why the fuel rods in a slow reactor are inserted in a water-filled chamber. Water acts a moderator material. The water-filled reactor chamber is sealed. The heat generated from the fission reactions causes the water to turn to high-pressure steam, which turns a turbine, which turns a generator to develop electrical energy.

A fast reactor uses a moderator material such as liquid sodium. The liquid sodium slows the velocity of neutrons, but not to the same extent as a water-filled reactor, hence the term—*fast reactor.* The fuel in a fast reactor contains some U-238 material. The faster moving neutrons impact the U-238 isotopes which absorb the neutrons and become plutonium isotopes (Pu-239) which in the wrong hands can be used in creating nuclear bombs and warheads. When the fuel rod material has given up its available energy, the plutonium portion can be reprocessed by mixing it with virgin U-235 material. This blended material is then re-inserted into fuel rods which are loaded back into the reactor.

In France, Germany, and Japan, the nuclear energy programs established decades ago were specifically structured around re-using some of the spent Pu-239 fuel material. Unfortunately, these nations are unique in this regard.

In 1977, U.S. bureaucrats made the decision not to mess with recycling of Pu-239. Instead, they opted to temporarily dispose of the waste by loading it into secure containers to be stored at various reactor sites across the nation. As of 2023, there are 77 storage sites spread across 35 states. These storage sites are home to an estimated 80,000 tonnes of radioactive spent fuel material. The plan was to one day have a central repository location somewhere in America that could safely hold all the waste. But that day has not come.

Yucca Mountain in Nevada at one time held high hope as a location to construct a central repository facility deep inside the mountain. Unfortunately, political wrangling and the NIMBY (not-in-my-back-yard) issue have prevented Yucca Mountain from being developed. People do not want anything nuclear located anywhere near where they live. In addition to the existing waste, the 93 operating reactors in the U.S. generate about 1,800 tonnes of additional spent fuel rods each year.

In 2021, talk turned to the concept of small modular reactors (SMRs) with annual power capacities ranging from as little as 10 MW to as much as 300 MW. A typical full-size reactor will generate around one gigawatt (GW) of power. There are several design iterations (slow and fast) of SMR being explored by countries around the world. The selling point of the SMR concept is that the surplus heat can be captured and used for processes like water desalination and industrial heating. Of course, the Second Law of Thermodynamics is not mentioned in any of these discussions.

People are generally buying into the SMR concept. The nuclear industry has done a masterful sales job and in late 2019 the Premiers of Ontario, New Brunswick, and Saskatchewan signed a memorandum of understanding in which they agreed to collaborate on the development and deployment of SMRs. The Province of Alberta joined into the collaboration in 2020. In 2021, Ontario Power Generation (OPG) announced plans to build a 300 MW unit at its existing Darlington reactor site by 2028. If this project gets off to a timely start, the other collaborating provinces could quickly follow suit.

Nuclear could be the answer to generating enough cheap electrical power to hydrolyze water into hydrogen that could be used in fuel cells in vehicles. Proponents of this idea argue that with an estimated 90 years of uranium resources left on the planet, uranium as a fuel source should be pursued. Detractors of the argument say no, because of the nuclear waste issue.

While a reactor does not generate carbon dioxide emissions, the issue of what to do with the existing nuclear waste (80,000 tonnes in the U.S. and 44,000 tonnes in Canada) and the new waste (1,800 tonnes created annually in the U.S. and an undisclosed amount created annually in Canada) must be reconciled with the fact that even SMRs will produce nuclear waste. There is no free lunch.

Finnish environmental company, Posiva Oy is in the final stages of construction of a series of underground tunnels carved out of the two-billion-year-old hardrock granite beneath Olkiluoto Island off Finland's west coast. Spent fuel rods from European reactors will be encased in a dual shell of iron and copper. The shells will be clad with a thick layer of bentonite clay before being placed in the tunnels, 420 meters below sea level. Plans call for this underground facility to hold 6,500 tonnes of spent fuel rods. While this sounds impressive, there will have to be many more facilities just like this one built around the globe to properly address the nuclear waste issue.

Canadian authorities have been watching the underground storage project in Finland. Sometime in 2024, the Nuclear Waste Management Organization (NWMO) will be announcing the location for a similar

CHARGED!

project in Canada. The location will either be near Ignace in north-western Ontario or in South Bruce County in southern Ontario.

- Nuclear reactors are no more or no less efficient that a coal-fired generating plant. The efficiency of a typical 1 GW reactor is about 40%.
- While nuclear plants do not produce carbon dioxide, the issue of nuclear waste is one that has been carefully concealed for decades now.
- Even the SMR reactor design now being touted by climate-change-fearing politicians will ultimately generate waste.

ELECTRICITY FROM THE SUN?

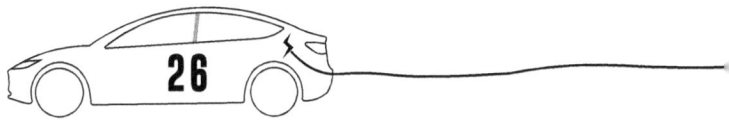

What if there was a heightened focus on generating electricity from solar panels? What if every vehicle charging station could be equipped with a solar panel array? What if homeowners could be incentivized to install solar panels on their rooftops to generate power to charge their electric vehicles while feeding any excess generated power to the electrical grid? What if solar panel electricity could be used to hydrolyze water to create hydrogen for fuel-cell powered vehicles?

Solar energy is a complex subject to navigate. Time of year, the geographic latitude of the solar panel location, weather conditions, ambient temperature, degree of tilt, and the general efficiency of the panels are all factors to consider. Plus, as discussed in an earlier chapter, materials like silver that are used in making solar panels is facing short supply.

The planets in our solar system orbit around the Sun following a plane of motion called the *ecliptic plane*. The Earth follows this ecliptic plane and orbits the Sun in 365 days. The Earth is tilted by an average of 23.5 degrees relative to the ecliptic plane. Because of this tilt, the amount of solar irradiation (solar energy) received by the northern hemisphere of the planet is greater between March and September. This explains why the northern hemisphere enjoys warmer weather generally from March (Spring Equinox) to September (Fall Equinox). The southern hemisphere enjoys warmer weather generally from September (Fall Equinox) to March (Spring Equinox) because of this tilt providing greater solar irradiation.

A solar panel in the northern hemisphere in the cooler months of September through March will still generate electrical energy, just less of it. For example, in Mossbank, Saskatchewan (where I live) a typical solar panel will generate about 5.79 kW·hrs per square meter of panel size per day in the period March through August. In the period September through March, that same panel will generate only about 2.66 kW·hrs per square meter of panel size per day.

The amount of solar irradiation hitting the Earth is always greater at locations nearer to the equator. At the equator, the Sun's rays hit the Earth at very nearly a 90-degree angle. This explains why a person near the equator at high noon will not see much of a shadow. At latitude locations further north, the Sun's angle of impact is less and the amount of solar irradiance that a solar panel will pick up diminishes. For example, solar panels on a building in Phoenix, Arizona will generate just over 7,000 kW·hrs per year. In Mossbank, Saskatchewan, where I live, those same panels would generate just under 4,900 kW·hrs per year.

At locations further away from the equator, the Sun will be at varying heights in the sky at high noon. This high-noon height will vary throughout the calendar year. In the autumn and winter months, the Sun will be relatively low in the noon sky. In June, at the longest day of the year, the Sun will be at its highest point in the sky at noon. A solar panel will therefore generate differing amounts of energy on a day-by-day basis during the year as a result of this changing solar angle. To partly compensate for this variability, a solar panel in a particular geographic location will have optimal angles of tilt that will help maximize the amount of solar energy obtained throughout a calendar year. For the summer season, this angle of tilt is given by the equation: tilt angle = (latitude x 0.9) + 29 degrees. In my case in Mossbank, Saskatchewan located at 49.9 degrees latitude, the tilt angle of a solar panel in the summer season, would be 73.9 degrees from vertical (16.1 degrees from horizontal). For the winter season, this angle of tilt is given by the equation: tilt angle = (latitude x 0.9) − 23.5 degrees. In my case in Mossbank, Saskatchewan located at 49.9 degrees latitude, the tilt angle of a solar panel in the winter would be 21.4 degrees from vertical (68.6 degrees from horizontal). These equations are overlooked by solar panel salespeople and their installers. The reason is, the angle of tilt of solar panels is not variable between seasons because the angle of the house roof does not change. Different styles of architecture will use roof pitch angles of perhaps 26 to maybe 40 degrees. As a result, a homeowner having solar panels installed is almost assured less than ideal electrical energy performance. The other aspect to consider is the fact that each day the Sun transits from east to west. A south facing roof will capture direct sunshine at noon. Prior to noon and after noon, the amount of irradiation absorbed will be less than optimal. Unless of course, the panels can be made to rotate their angle during the day. This technical accommodation is not a feature of standard solar panels for houses.

In order for a solar panel to create electrical energy, rays from the Sun must hit the panel. There is no Sun at night, and on a cloudy day the

Sun might not be seen through the overcast conditions. Furthermore, electricity generation will be seriously reduced if the panel is covered with a layer of ice, or snow in winter. For many locations in North America, solar panels are not the dream solution that consumers have been led to believe.

Another variable to consider is the efficiency of the solar panels. A solar panel is comprised of two silicon wafers sandwiched on top of one another. The wafers are assembled into modules and the modules in turn are assembled into the panels that you see on residential house roofs. The top silicon wafer has been impregnated (doped) with atoms of phosphorous. The atoms of phosphorous make the electrons in the silicon crystalline structure unstable and the wafer layer will seek to rid itself of electrons. The bottom wafer has been impregnated (doped) with atoms of rare elements such tellurium or gallium which render the crystalline structure of this layer amenable to receiving electrons. One layer seeking to rid itself of electrons and one layer seeking to accept electrons makes possible the *photovoltaic effect*. When photon energy from the Sun hits the top wafer layer, the photon energy boosts electrons out of their orbital shells and sends them towards the lower wafer where they are accepted. This flow of electrons is by definition— electricity.

By doping the silicon structure with elements other than just phosphorous, tellurium, or gallium, solar panel designers can adjust the *band-gap*. This refers to the amount of solar photon energy needed to boost an electron out of its orbital shell and send it hurtling towards the lower wafer. Solar panels are engineered with a band gap of 1.1 to 1.7 electron volts (eV). The entire spectrum of light coming from the Sun, will vary in wavelength depending on the time of day. The energy of the Sun's photons will cover a range of about 0.5 eV to about 2.9 eV. This means that early in the morning or late in the evening when the Sun's photons exhibit energy outside the band gap of the panel structure, there will not be sufficient photon energy to create much electrical energy.

In 1961, scientists William Shockley and Hans Queisser realized that the portion of the Sun's photon energy greater than the band gap of the silicon wafer in the solar panel was wasted as heat. This led them to arrive at the Shockley-Queisser maximum theoretical efficiency limit for silicon solar panels. This efficiency limit is a mere 33.7%, and that of course under ideal conditions. Move to a more northerly latitude, encounter changing seasons, rain, clouds, snow, or adverse weather, and the efficiency of a panel lessens notably.

A typical solar panel (5 feet x 3 feet) has a capacity of near 300 Watts. Suppose that six of these panels were assembled together on the roof of a house. Suppose that the Sun provided good amounts of solar radiation 300 days (10 months) out of the year for 12 hours per day. The remaining 65 days are either cloudy or have ice/snow buildup on the panels. In theory, this array of panels would generate 6,480 kW·hrs of energy in the 300 days (21.6 kW·hrs per day). Now adjust for the reality of Shockley-Queisser efficiency and this array will generate 2,140 kW·hrs (approximately 214 Kw·hrs per month). In a typical month, I consume nearly 575 kW·hrs at my house to power lights and appliances. This 5 foot x 3 foot panel would only provide me with about 1/3 of the monthly electrical energy I need. Suppose I wanted to also charge an electric vehicle. Suppose this vehicle had a 100 kW·hrs battery pack and that I would only ever charge it to 80% capacity (80 kW·hrs). This solar panel does not even provide enough power to run my house. Adding an electric vehicle into the mix is out of the question. To have enough solar panel to run my house and to charge an electric vehicle once a week that I was driving 80 kms (50 miles) per day, would mean having to install five of these 5 foot x 3 foot panels on my rooftop. Two of these panels would be dedicated to just providing the electricity to charge my vehicle. If I was driving 160 kms (100 miles) each day from my rural location to a nearby city for work, this would mean installing seven panels on my rooftop. Four of these would be dedicated just to charging my vehicle.

The notion of equipping a rapid charging station with solar panels is even further out of the question. A rapid charging station will provide around 80 kW·hrs of energy to a vehicle battery pack in 20 minutes. It would take over 80 of these 5 foot x 3 foot panels to create enough electrical energy to allow a rapid charging station to charge two vehicles per hour every day during daytime hours when the Sun is shining.

In other words, solar energy is not the answer to a population of consumers seeking to charge their vehicles. Solar panels are not efficient enough. This inefficiency further destroys the argument that solar power might be an answer to hydrolyzing water to make hydrogen fuel. A 5 foot x 3 foot solar panel will generate enough energy to hydrolyze only three liters of water each day.

- Despite the popularity of solar panels, solar is an inefficient way of producing electrical energy for both homeowners and electric vehicle owners. There is no free lunch.

FOOD FOR FUEL?

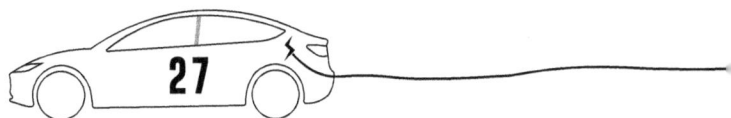

27

Farmers in the U.S. mid-west grow corn. Farmers in Canada grow wheat. Using these grains as feedstock, each year North American ethanol plants produce approximately 15 billion gallons (56 billion liters) of ethanol (C_2H_5OH). In 2018, I had a tour of the ethanol production facility in the small town of Unity, Saskatchewan. This facility was efficiently making ethanol from wheat 24 hours a day, seven days a week. I was impressed.

Ethanol is already blended into automotive fuels. The next time you are filling your vehicle with gasoline, look at the pump and you will see a sticker bearing a message that the fuel contains about 10% ethanol.

The bigger question is: could ethanol be used in vehicles equipped with fuel cells? The short answer to this question is—yes. Ethanol could be used as a fuel in Direct Ethanol Fuel Cells (DEFC). The engineering challenge would be to fit the anode and cathode sides of the fuel cell with suitable catalyst material to drive the electricity-making chemical reactions. Italian firm Acta S.p.A. is making strides with its *Hypermec* catalysts for Direct Ethanol Fuel Cells. It has developed a polymer material impregnated with iron, cobalt, and nickel.

At the anode side of an ethanol fuel cell, the reaction would be:

$$C_2H_5OH + 3\ H_2O \rightarrow 12\ H^+ + 12\ \text{electrons} + 2\ CO_2.$$

At the cathode side of the cell the reaction would be:

$$3\ O_2 + 12\ H^+ + 12\ \text{electrons} \rightarrow 6\ H_2O.$$

The downside to Acta's fuel cell design is its 40-45% operating efficiency. This brings thermodynamics back into focus. It takes energy to make energy.

Proponents of biomass as an energy source typically look at a bushel of corn or wheat and imagine it being effortlessly converted into fuel. What they do not consider is the various steps involved in creating the corn or wheat and the steps in processing it into ethanol.

A farmer will consume fuel to move his machinery through the field at seeding time. He will add various amounts of fertilizer to the soil, all of which required energy to produce. Later in the growing season, he will incur fuel expenses as he drives his sprayer through his fields to apply pesticides and herbicides to the land. These chemicals all required energy to produce. He will incur further fuel expenses at harvest time as he uses his combine to harvest the crop. The harvested grain will then be trucked to an ethanol plant. Energy will be consumed at the plant to process the grain and distill it into ethanol. The waste grain mass left over will then be dried in a kiln before being delivered to a cattle feedlot operation. The entire production process from start to finish will thus create carbon dioxide.

There are several engineering studies that focus on the energy balance of turning corn into ethanol. These studies show that the energy input is around 74,500 British thermal units (Btu) per gallon of ethanol made. Consider that the energy contained in one gallon of ethanol is around 76,000 Btu. Therefore, ethanol from corn is only marginally beneficial. Yet the politicians in Washington continue to encourage the use of corn for ethanol through a complex web of tax subsidies to farmers. Would it not make more sense to use this corn to feed the world?

Corn contains molecules of glucose which is a fermentable sugar. The fermentation reaction is:

$$C_6H_{12}O_6 \rightarrow 2\ C_2H_5OH + 2\ CO_2 + heat.$$

These equations show that one mol of glucose will generate two mols of carbon dioxide and two mols of ethanol. In other words, 180 grams of glucose will generate 88 grams of carbon dioxide and 92 grams of ethanol.

The U.S. government mandates that a gallon of gasoline must contain 10% ethanol. In 2023, the mandated volume of ethanol that was used in the U.S. was near 21 billion gallons (79.5 billion liters), based on the amount of gasoline used across the nation.

Consider that corn contains around 62% glucose, with the balance being protein, water, and fiber. This means it will take 290 grams of corn to provide the 180 grams of glucose in the above equation. The density of ethanol is 0.789 kg/liter. The 92 grams of ethanol in the above equation will occupy a volume of 0.1166 liters. To make 79.5 billion liters of ethanol will require 198 billion kgs of corn (7.77 billion bushels). At a typical corn yield of 160 bushels per acre, it will take nearly 49 million acres of farmland to produce the corn to make the mandated ethanol. *All* of the farmland in Iowa *and* Indiana combined only amounts to 45 million acres. Farmers absolutely must grow corn in order to produce the mandated amounts of ethanol. This explains the maze of tax subsidies and financial enticements.

A further issue with making biofuel is emissions. The above equation shows that 180 grams of glucose will release 88 grams of carbon dioxide. This means that the 198 billion kgs of corn processed will release nearly 97 million tonnes of carbon dioxide. Furthermore, the fermentation process for 198 billion kgs of corn will consume near 700 billion liters (185 billion gallons) of water.

And then there is the issue of emissions from combustion. When ethanol is combusted as a fuel, one mol (46 grams) of ethanol will generate two mols (88 grams) of carbon dioxide. The chemical reaction is: $C_2H_5OH + 3\,O_2 \rightarrow 2\,CO_2 + 3\,H_2O + heat$. The 79.5 billion liters of ethanol consumed each year in the U.S. will release nearly 120 million tonnes of carbon dioxide.

Wrapped inside these technical explanations is the reality that corn is food for the global population and this food is being used to make fuel that has only marginal energy benefits.

- It takes energy to make energy. Making energy can release carbon dioxide into the environment. Corn-based ethanol used each year in the U.S. releases over 200 million tonnes of carbon dioxide into the atmosphere. Even if this ethanol could be efficiently used in ethanol fuel cells to power vehicles, the marginal efficiency and the carbon dioxide generation of the ethanol process still remain as issues.

RECYCLING OF BATTERY METALS

28

As a dim flashlight reminds you, battery systems do not last forever. The same applies to electric vehicle batteries. At some point spent vehicle batteries must be disposed of. The *E.U. Battery Directive 2006/66* issued in 2006 mandated that E.U. member states take steps to recycle batteries. The level of urgency to begin recycling batteries was increased in late 2022 with the signing of the *European Green Deal*. Under this deal, the lithium from non-vehicle batteries must be 63% recovered through recycling by 2027 and 73% recovered by 2030. The lithium from vehicle batteries must be 51% recovered by recycling by 2028 and 61% recovered by recycled by 2031. Other metals such as cobalt, nickel, and manganese face recovery targets of near 95%.

Meanwhile in the U.S., politicians are doing what they do best—talking. The U.S. Department of Energy estimates that by 2023 there will be near 165,000 vehicle batteries in need of recycling. The closest the U.S. has come to mandating action on battery recycling is the *2022 Strategic EV Management Act* which directs federal agencies to work towards guidelines for recycling batteries from government electric vehicles.

More work has to be done to better define protocols and procedures for battery recycling. E.U. legislation (that essentially kicks the can down the road to 2028) and no clearly defined federal battery recycling legislation in the U.S. mean that battery recycling is at a standstill for now.

However, while the political class dithers, academia has released dozens of studies investigating different processes for recovering the cathode battery materials. A 2022 review paper notes that NCA and NCM battery designs are suited to hydrometallurgical techniques for recovery of the cathode metals. $LiFePO_4$ (LFP) battery designs are best suited to pyrometallurgical techniques. There is no generic recycling technique for all batteries.

Hydrometallurgical processes involve crushing and finely grinding materials which are then separated from each other using an aqueous media of water and chemicals. Pyrometallurgical processes involve crushing and finely grinding materials which are then smelted at high temperatures. In the molten state, the materials will separate from each other based on their densities.

Academic literature suggests that a considerable amount of work has been done at institutions around the world to explore hydrometallurgical battery recycling procedures. One challenge that exists is with the crushing of the spent batteries to remove the outer casement material and to remove the thin foils of aluminum and copper current collector material. This involves human labor which costs money. Another challenge comes with selectively being able to isolate both cobalt and nickel in the hydrometallurgical electrochemical process. Another issue that arises is the harsh nature of the various chemicals and reducing agents used in the process. Where will residual process sludge be disposed of? What are the safety risks to employees when dealing with these chemicals?

A considerable amount of work has also been done to explore pyrometallurgical techniques for recovering metals from LFP batteries. A 2021 study by scientists in Austria showed that a pyrometallurgical approach of heating the cathode material to near 1400°C achieved a lithium removal rate of 68.4% with a phosphorus removal rate of 64.5%.

The private sector has used this information from academia and taken steps to pursue the recycling of battery metals. As of late 2022, there were at least 32 established or planned facilities around the globe for lithium-ion battery recycling. Collectively, these facilities offer or will offer roughly 400,000 tons of annual recycling capacity. This is assuming that the economics of recycling are robust enough to attract investment capital.

An example of a company that has run headfirst into the economics of battery recycling is Canadian-based Li-Cycle Holdings (shares trade on the New York exchange as LICY). The company website explains that its business model is focused on using its patented process to recover the battery metals from all sizes of spent lithium-ion batteries. Few process details are provided on the website.

What is not made clear is how the company will obtain spent vehicle batteries from consumers. Will car dealers collect the spent batteries? What will transporting the batteries to a recycling depot cost? Financial results for Q3 2023 appear to shed some light on the story. For Q3, the company declared an accounting loss of over US$130 million as it burned through $150 million in cash. As of early March 2024, LICY shares were trading at below 40 cents per share. The New York Stock Exchange has issued a warning that the share price must be brought back above the $1 per share level if the exchange is to continue listing the shares.

The State of California is adding leverage to the battery recycling story through *Senate Bill 615* which mandates: *by January 1, 2025, vehicle makers, battery manufacturers, and repair shops are jointly responsible to collect the depleted battery, repurpose it if possible, and ensure that it is recycled if it cannot be repurposed.* As of late January 2024, this Bill is at the committee stage and is awaiting its first reading in the California legislature. Whether this Bill will gain momentum remains to be seen. If this Bill can make headway, investment capital might find its way to companies struggling to establish recycling processes.

- Recycling of EV lithium-ion batteries will not be a straightforward approach. It might not even be economically feasible. Safety, waste disposal, and process efficiency are all issues that will all have to be addressed.

WHAT IS NOT BEING TALKED ABOUT

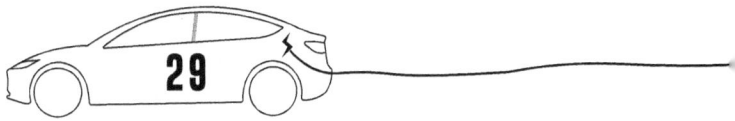

This book has covered considerable ground in exploring the electric vehicle theme. Behind the scientific and environmental issues that have been discussed, there are economic and safety issues to be aware of as well. Many of these issues are not covered by media.

THE AUTOMAKERS ARE LOSING MONEY

To use an old adage, the automakers are between a rock and a hard place. They are being mandated by way of ZEV credits and fleet emissions enticements to manufacture more electric vehicles. But the move towards electric vehicles is proving to be a costly proposition that is weighing on company share prices and incurring frustration among stock market analysts.

Retooling assembly lines to produce a radically new design of vehicle is not an easy task, especially if the workers on that assembly line have only ever produced vehicles with internal combustion engines. In the first quarter (Q1) of 2023, Ford delivered 12,000 electric vehicles to dealerships. Statements filed with the Securities and Exchange Commission (SEC) show a Q1 loss of $722 million. This equates to a loss of near $60,000 for each electric vehicle. For the second quarter (Q2) of 2023, Ford delivered 35,000 electric vehicles to dealerships. SEC filings show a loss of $1.1 billion which equates to a loss of near $35,000 per vehicle. For the third quarter (Q3) of 2023, Ford delivered 35,000 electric vehicles to dealerships. SEC filings show a loss of $1.3 billion which equates to a loss of near $37,000 per vehicle. As these electric vehicle losses continue to mount, it is gasoline powered vehicles and vehicle financing that are generating the profits for Ford. To provide more clarity to investors and analysts, in mid-2022, the company reorganized itself into three operating divisions: the Model E division (electric vehicles), the Model Blue division (internal combustion engine vehicles), and Ford Pro (sales, distribution, and service). I would not be surprised to see the company go one step further and create a separate publicly-traded entity for the electric car division to safeguard its core vehicle brands. Company management is not going to sit idly by as analysts downgrade Ford share price targets. Why imperil the entire ship over one piece of bad cargo?

The situation at General Motors is similar. The sales of gasoline powered vehicles are funding the losses of the electric car segment. As of mid-2023, the company offered an electric Hummer, the Cadillac Lyriq, and two models of Chevy Bolt. Several models of pickup trucks will soon be available in electric format. To minimize the losses going forward, General Motors is telling stock analysts that it plans to target the costs of batteries. Its planned next-generation Ultium battery design will be 60% cheaper than the batteries used in the Chevy Bolt and 40% cheaper than the current design of Ultium battery. The company has offered little in the way of detail as to how it will slash these costs. Given the 2021 recall issues encountered with Chevy Bolt batteries, one would think that opting for a cheaper battery design would be the last thing the company would do.

Automaker Stellantis (formerly Chrysler) does not mince words. Stellantis is further along the development path than Ford and General Motors. Globally, it currently offers 19 electric models and 15 plug-in hybrid models. However, the company states that it costs 40-50% more to build an electric vehicle than a gasoline-powered vehicle. This additional cost cannot be absorbed by the company, nor can it be passed on to the consumer in the form of a higher vehicle price. The company states that it plans to bargain harder with parts suppliers to reduce the overall costs of making an electric vehicle.

Apple Scraps Project Titan

Speaking at a conference sponsored by the *Wall Street Journal* in 2015, Apple CEO Tim Cook took aim at the car industry. He explained how he felt the car industry was at an inflection point for massive change. He explained that the shift from internal combustion engines to electric vehicles was being driven by the use of software in cars and by efforts to create autonomous vehicles. What Cook did not elaborate on was that Apple had already hired several thousand employees to work on vehicle projects. However, Project Titan is now dead and buried. In February 2024, Apple abruptly announced that it had abandoned its vehicle development efforts. The employees working on the project would be reassigned to artificial intelligence (AI) projects. When Apple gives up on a project after eight years, you know that something is wrong.

THE WALLS ARE GOING UP

The Apple announcement came at a curious time. Several days after the Apple announcement, the White House issued a press release quoting President Biden who said, "China is determined to dominate the future of the auto market, by using unfair practices. China's policies could flood our market with its vehicles, posing risks to our national security. I'm not going to let that happen on my watch."

This press release opens the door to deeper investigations on the risks posed by foreign vehicles on U.S. roadways. The fear is not the fact that Chinese vehicles from companies like BYD or Nio are electric. As Commerce Secretary Gina Raimondo noted, the fear is that the vehicles are connected to the internet. She said they are "like smart phones on wheels. They collect huge amounts of sensitive data on the drivers—personal information, biometric information, where the car goes. So it doesn't take a lot of imagination to figure out how a foreign adversary like China, with access to this sort of information at scale, could pose a serious risk to our national security and the privacy of U.S. citizens."

Is this reaction really about security? Or is this reaction providing cover for the fact that General Motors, Ford, and Stellantis are all struggling to find profitability in the electric vehicle segment? Is this reaction a tactic to garner votes from unionized auto workers? The U.S. already applies a 25% tariff to imported Chinese electric vehicles. If a Chinese vehicle is partly assembled in Mexico and then sent to the U.S., there is an additional 2.5% tariff. If Missouri Senator Josh Hawley gets his way, these tariffs will increase even further to 125%.

M.G. BUCHOLTZ

COLLISION SAFETY

The weight of the battery pack in a typical electric vehicle is between 300 and 600 kgs, depending on the energy content of the battery. Consider that an internal combustion engine to power similar sized cars will weigh around 200 kgs. To provide for energy efficiency and keep weight down, automakers have resorted to lighter-weight polymer materials in the construction of electric vehicles. Despite these efforts, a typical 4-door electric vehicle will tip the scales at 2,300 kgs. An average full-size car with an internal combustion engine will weigh around 2,000 kgs. An average mid-size car with an internal combustion engine will weight around 1,500 kgs.

The question to be asked is: how structurally safe is a vehicle that has been redesigned with light-weight materials so as to accommodate the heavy battery pack? The kinetic energy of a moving object impacting another object is given by the formula ½ mv². A sedan weighing 1,800 kgs travelling at 100 km/hr (27.7 m/sec) will have an impact kinetic energy of 691 kJ. An electric vehicle weighing 2,300 kgs travelling at 100 km/hr (27.7 m/sec) will have an impact kinetic energy of 882 kJ. In a collision scenario, more damage will be done; the National Bureau of Economic Research estimates a 47% greater probability of fatality if impacted by a vehicle weighing 1,000 kgs more than the one you are driving.

VEHICLE WEIGHT

What about parking garages in cities? Many of these structures were built decades ago and are not designed to accommodate extra weight. What about the extra weight of the electric vehicles travelling on roadways? How much incremental wear and tear is exacted on paved surfaces as a result of the heavier vehicles? In the U.S., drivers fueling up at a gas station pays taxes of between 18.4 cents and 65 cents per gallon, depending on the state. Some of this tax is usually directed towards efforts to maintain roadways across the nation each year. In Canada, the fuel taxes levied at the pump are much higher—at around 46 cents per liter ($1.79 per gallon). Electric vehicle drivers obviously avoid this tax, at least until recently. In March 2024, the Alberta government announced that starting January 1, 2025 it will levy a $200 per year tax on owners of electric vehicles. The howls of derision from electric vehicle owners were deafening. However, electric vehicles are causing wear and tear on the roads and it is only fair that the owners of these vehicles pay their fair share.

THE ECONOMICS OF BATTERY CHARGING

As more and more supercharger stations appear at service stations and other locations, rumors continue to mount about the merits of using these fast chargers which have the ability to replenish a battery to 80% of full charge in less than 30 minutes.

There are three types of charging station available for electric vehicle owners: Level 1, Level 2, and Level 3. The first two types draw alternating current (AC) at 120 volts (12 amps) and 240 volts (up to 80 amps) respectively and then use the inverter on the vehicle's built-in charger apparatus to convert the AC current to direct current (DC) so that the battery can accept the charge. Level 3 chargers provide DC power directly to the battery at up to 480 volts and 300 amps which means a shorter charge time.

Lithium-ion batteries should ideally be fast-charged at between 5°C to 45°C. Attempting to charge the battery below 0°C (32°F) will result in lithium ions not properly intercalating into the anode and instead building up on the anode surface. This will lead to permanent battery degradation.

This does not bode well for car owners in cold climates who do have garages.

Fast charging stations employ high-voltage to charge the battery. As the charge is applied, lithium ions are forced from the cathode and moved to the anode. The higher the voltage, the higher the force that moves the lithium ions. Fast charging has the potential to cause cracks in the cathode which become sites for lithium ions to accumulate. The expression *dendrites* refers to lithium buildup sites. Cracks and dendrites will reduce the capacity of the battery.

In 2014, researchers at the Idaho National Laboratory used four 2012 model year Nissan Leaf electric vehicles in a charging mode test. The cars were driven extensively around the city of Phoenix, Arizona. Two of the cars were recharged at fast charging stations. The other two were recharged at Level 2 charging stations. At the beginning, the battery packs were tested every 10,000 miles. The vehicles charged with the Level 2 chargers exhibited a battery capacity of 23.45 kW·hrs. After 50,000 miles and numerous recharges, the capacity had diminished to 17.64 kW·hrs. The vehicles charged with the fast chargers exhibited an initial battery capacity of 23.31 kW·hrs. After 50,000 miles and numerous recharges, the capacity had diminished to 16.93 kW·hrs.

The conclusion was that remaining battery capacity figures differed only by 4%, therefore fast charging was not detrimental to batteries. Although the study said that fast charging was not detrimental to the battery, a capacity decline from 23.45 kW·hrs to 17.64 kW·hrs is a 24.7% deterioration after only 50,000 miles (80,400 kms). The key takeaway for the consumer is that, based on this Nissan Leaf study, repeated and sustained use of fast charging will lead to reduced energy storage capacity, a shorter range, and a higher frequency of battery changes over time. Not exactly stellar selling features.

DRIVING RANGE

Driving range of an electric vehicle tends to be first and foremost on the mind of a consumer. An online search on the subject will produce a plethora of websites and blogs focused on driving range. The following table presents some figures from various automaker websites.

Model	Battery Size	Driving Range
Tesla Model S	100 kW·hr	405 miles (652 kms)
Ford Mustang Mach E	99 kW·hr	217 miles (350 kms)
Tesla Model 3	89 kW·hr	272 miles (437 kms)
BMW iX	80 kW·hr	324 miles (521 kms)
Mercedes EQA	79.8 kW·hr	350 miles (563 kms)
Hyundai Ioniq	77.4 kW·hr	361 miles (581 kms)

What is not discussed is how these numbers are calculated.

In 2007, the E.U. set out to establish a harmonized testing procedure to determine the driving range of electric vehicles. After many years of study, in 2018 a procedure was finally enshrined as *Regulation (E.U.) 2018/1832*.

The test procedure is based on three different driving cycles which represent three different power to mass (PMR) ratios of vehicle. The PMR is defined as the ratio of the vehicle's rated power (W) divided by its mass (kg).

Class 3 vehicles (PMR greater than 34) are typical of those that E.U. consumers would own. The test conditions were determined from real-world driving data obtained from five different regions: the E.U. (including Switzerland), the U.S., India, Korea, and Japan.

Test vehicles were mounted on a dynamometer—essentially a treadmill for cars. The dynamometer consists of rollers to which the vehicle is secured. The vehicle can then be placed in drive and the test technician can control the braking and the rotation of the rollers.

Class 3 vehicles are subjected to:

- a 590 second drive at a maximum speed of 56.5 km/hr, followed by a 145 second stop.
- a 433 second drive at a maximum speed of 76.6 km/hr, followed by a 47 second stop.
- a 455 second drive at a maximum speed of 97.4 km/hr, followed by a 29 second stop.
- a 323 second drive at a maximum speed of 131.3 km/hr, followed by a 6 second stop.

In total, a test would have the vehicles wheels cover 23.2 kms in 1,801 seconds interspersed with 227 seconds of stoppage. The temperature in the testing laboratory was fixed at 23°C. The amount of electricity to bring the battery pack back to full charge was then recorded and the drive range estimated.

In the U.S., electric vehicle range tests are conducted by the Environmental Protection Agency (EPA) also using a dynamometer. The EPA has three test cycles that it subjects vehicles to: the *Urban Dynamometer Driving Schedule* (UDDS), the *Highway Fuel*

Economy Driving Schedule (HWFET), and the *Multi-Cycle City/Highway Schedule.*

The UDDS cycle simulates stop-and-go city driving. The vehicle is taken from a standstill to an average of about 35 miles per hour (56 kms/hr) and back to a standstill multiple times over a test period of 1,352 seconds.

The HWFET cycle simulates sustained-speed highway driving. The vehicle is taken from a standstill to an average speed of 48 miles per hour (77 km/hr). The speed fluctuates frequently between 30 and 60 miles per hour over a test time of 765 seconds.

The Multi-Cycle City/Highway cycle involves fully charging the vehicle and leaving it parked overnight. The following morning, the vehicle is mounted on the dynamometer then put through multiple UDDS and HWFET cycles until the battery is completely discharged and the car can no longer drive. The car is then plugged in with the charger provided by the automaker and charged back to full.

To compensate for the idealistic nature of the tests, the EPA multiplies the range numbers from the tests by 0.7 to provide a final range that is more in line with what drivers can expect from their cars under real-world conditions. The mathematical calculations use 33.705 kW·hrs of electricity as being equivalent to the energy from one U.S. gallon (3.78 liters) of gasoline.

Since the 1950s, *Car and Driver* magazine has been subjecting vehicles to real-time, on-road city and highway fuel-economy tests. Over the past couple years, the magazine has arranged for real-time road testing studies on various models of electric vehicles. The authors of these studies have concluded that most of the electric vehicles tested have fallen short of both their EPA electric consumption and driving ranges. These studies noted that the discrepancy between EPA numbers and real-world numbers is exacerbated by temperature, road surface conditions, regenerative braking, and the degree of charge the battery is taken to. The studies concluded the EPA must re-evaluate its testing procedures to better reflect reality.

TESLA DRIVING RANGE CONTROVERSY

On July 28, 2023, the internet erupted with controversy when Reuters News released a report claiming that Tesla was deliberately misleading consumers over driving range. The article was based on information given to the Reuters journalists by an unnamed source.

The source pointed the journalists to a person who had a less-than-pleasant driving experience in early 2023. The journalists located the person in question, Alexandre Ponsin, and the story began to take shape. After purchasing a used 2021 Model 3 vehicle, Ponsin set out on a family road trip from Colorado to California. He expected to get something close to the electric sport sedan's advertised driving range: 353 miles on a fully charged battery. He soon realized he was sometimes getting less than half that much range.

Ponsin described how after arriving in California, he contacted Tesla and booked a service appointment at a nearby dealership. He soon received two text messages: one telling him that remote diagnostics had determined his battery was fine, the second message reading: *We would like to cancel your visit.*

The journalists kept digging and what they found next was shocking; Tesla employees had been instructed to prevent any customers complaining about poor driving range from bringing their vehicles into dealerships. In 2023, Tesla had created a "Diversion Team" in Las Vegas to cancel as many range-related complaint appointments as possible. Tesla was finding that its service centers were inundated with appointments from owners who had expected better performance based on the company's advertised estimates and the remaining range projections displayed by the in-dash range meters of the cars themselves. The Team often closed hundreds of cases a week and Tesla call center employees were tracked on their average number of diverted appointments per day.

It was determined that Tesla had been exaggerating its vehicles' potential driving distance for years, using software algorithms. The in-car dash display would present the driver with a favorable-looking projection for the distance the vehicle could travel on a full battery. Then, when the battery fell below 50% of its maximum charge, the algorithm would show a more realistic projection of the remaining driving range. To prevent drivers from getting stranded as their predicted range started declining more quickly, the vehicle had been equipped with a "safety buffer," allowing about 15 miles (24 km) of additional range even after the dash readout showed an empty battery. As Mr. Ponsin described, he would be looking at the remaining range on the dash readout and he would see the number rapidly decrease in front of his eyes.

The Reuters journalists went on to discover that in early 2023, Tesla had been fined $2.2 million by the South Korean Fair Trade Commission. Regulators found the cars delivered only up to 50% of their advertised range in cold weather.

What is interesting about the Korean fine and the Reuters article is that Tesla made no effort to engage the media with an explanation. Silence was the response.

In 2023, Gregory Pannone, Director of Fuel Economy and Performance at the *Society for Automotive Engineers* (SAE), co-authored a study of 21 different brands of electric vehicles. The report revealed that, on average, the cars tested for the report fell short of their advertised ranges by 12.5% (under highway driving conditions) and that three Tesla models posted the worst performance of all the vehicles tested, falling short of their advertised ranges by an average of 26%.

INSURANCE AND REGISTRATION

Data obtained from Saskatchewan Government Insurance (SGI) raises another troublesome issue regarding electric vehicles. A collision involving an electric vehicle could cause significant damage to the complexities of the battery system. For this reason, the insurance cost of electric vehicle is higher than for gasoline powered vehicles. To cite an actual example, the insurance cost for a Tesla, Model 3 is $2,023 per year; the insurance cost of a gasoline-powered Hyundai Sonata is $1,481. In addition, there will be a $150 surcharge fee levied when a Tesla owner registers the vehicle and obtains the license plates.

An article that appeared on Bloomberg in late January 2024 pointed out that increased insurance rates on electric vehicles have now reached the U.K. The average insurance premium for electric vehicles jumped to 1,344 Pounds ($1,700) at the end of 2023. The cost of insuring electric vehicles in the U.K. is now double the cost of that for traditional cars. Commenting on this increase, U.K. insurance broker, Howden Group Holdings pointed out that higher repair costs, more time spent in workshops, and a lack of mechanics trained to fix electric vehicle batteries are all factors in the premium increase.

GENERAL MOTORS KICKS THE EV DOWN THE ROAD

As I was preparing to hand this manuscript over to my publisher, the sands under the electric vehicle argument suddenly shifted yet again. On January 30, 2024, General Motors CEO Mary Barra issued a press release that showed an obvious level of discomfort with fully-electric vehicles. She explained to stock market analysts that General Motors plans to reintroduce gas-electric hybrid vehicle models while staying committed to ramping up production of battery-only vehicles by 2035. To hear the CEO of a major vehicle maker suddenly have a bad case of buyer's remorse was another bit of evidence supporting that the arguments made in this book.

WHAT IS BEING SAID IN WASHINGTON

In June 2023, the Subcommittee on Economic Growth, Energy Policy, and Regulatory Affairs convened a hearing. This subcommittee is a subset of the larger Committee on Oversight and Responsibility.

In one of the exchanges, Lauren Boebert (Republican, 3rd Congressional District of Colorado) asked questions of Joseph Goffman, the Deputy Administrator of the EPA's Office of Air and Radiation. Under questioning from Ms. Boebert, Mr. Goffman admitted he did not know how the average electric vehicle price compared to the average gasoline-powered vehicle, except to say he thought electric vehicles might be more expensive.

In another exchange, Byron Donald (Republican, 19th Congressional District of Florida) and Pat Fallon (Republican 4th Congressional District of Texas) asked questions of Stephen Bradbury. Mr. Bradbury is a Fellow at the Heritage Foundation. Prior to taking a position at this think-tank, he was the General Counsel to the U.S. Department of

Transportation (DOT) in the Trump Administration. In the exchanges with Mr. Donald and Mr. Fallon, Bradbury made it clear that the average electric vehicle will require six times more mineral inputs than a conventional vehicle. He stated that if 50% of vehicles on U.S. roadways today were electric, the electric grid would not be able to function. He further stated that if every country in the world achieved its electric vehicle targets, the total savings in CO_2 emissions would reduce global temperatures by a mere 0.002 °F by the year 2100.

What is more interesting is that Mr. Bradbury came to the hearing armed with a 21-page report. His report takes issue with how the Biden-appointed Environmental Protection Agency (EPA) bureaucrats are usurping too much authority. The following excerpts are taken from his report:

... *because Congress has recently approved generous federal subsidies for some EV purchases and charging infrastructure, the EPA says it can now declare that battery-electric vehicle technology is a "feasible" alternative to the traditional internal-combustion engine (ICE) powertrain.*

... *on that basis, the EPA is proposing to treat EVs as an available "control technology" for achieving compliance with the tailpipe emissions restrictions under section 202 of the Clean Air Act.*

...*the automakers are pledging to invest in the transition to EVs because governments around the world—like China, the EU, the Biden White House, and Governor Gavin Newsom and his climate regulators in California—are demanding that they do so. But everyone knows there is a large looming impediment to this Green Dream: resistance from American consumers. The American public is not jumping on the electric bandwagon. EVs are expensive—beyond the reach of many American families—and most Americans remain skeptical that EVs will reliably serve the full range of their needs, that quick and convenient charging stations will be widely available, that EVs will maintain their promised driving range over time or in cold weather, that they will have any resale or trade-in value whatsoever, and that insurance carriers will cover the huge costs of battery replacement when the battery is damaged in a minor accident.*

To push the automakers to convert to EV production in the absence of sufficient market demand, EPA plans to ratchet down the emissions limits for carbon dioxide and for other pollutants associated with smog (such as unburned hydrocarbons, particulate matter, oxides of nitrogen, and ozone) to super-stringent levels that are technologically impossible for gas-powered vehicles to satisfy.

Indeed, the current proposals represent an extreme example of regulatory overreach. The EPA sees an endless horizon for its new-found power to regulate practically all aspects of the American automotive market.

And if there truly were an explosion in the sale of EVs, those EVs would need to be charged using electricity produced mostly from fossil-fuel-fired power plants, increasing the national emissions of carbon dioxide. EPA largely dismisses this reality based on the wishful claim that America's future power generation will soon shift en-masse to wind and solar.

To accommodate EVs, our national electric grid capacity would need to grow 60 percent or so by 2030 and much more over the long term, and that is growth in infrastructure alone, not in power generation. This buildout will have to be paid for, and those costs will inevitably be reflected in higher electricity rates and higher EV charging fees. EPA says not to worry about grid reliability—because the government will be able to manage the EV charging draw on the grid by rationing the hours for charging. American drivers will not tolerate that.

Driving a single EV 15,000 miles per year and charging it at home could raise the annual electricity bill for the average family by 50 percent or more. If the nation converts to EV ownership at the rates EPA is aiming for, where is the additional electricity needed to power those vehicles going to come from? How could such a large increase in the draw on the grid not cause electricity rates to rise significantly?

China controls nearly 70 percent of global EV battery manufacturing capacity—including 70 percent of the world's lithium supply; 80 percent of the necessary rare earth minerals; and approximately 75 percent of the magnets needed for EV motors—and it boasts 107 of the 142 lithium-ion battery mega-factories planned or under construction in the world today (with only 9 planned for the U.S.) and represents far and away the largest EV market.

The average EV battery uses about 8-10 kilograms of lithium (even more for higher performance batteries), and the world today mines a total of about 130,000 tons of lithium per year. That means if the EPA succeeds in converting 60 percent of annual U.S. car sales to EVs (about 7.8 million vehicles), those EVs (just for the U.S. market) would require 60 percent of the entire world's current production of lithium. Similarly, each EV battery requires about 10 kilograms of cobalt, which translates into 1 metric ton for each 100 EVs and 10,000 tons of cobalt for 1 million new EVs. There are only between 150,000 and 190,000 tons of cobalt mined every year worldwide (the lion's share from the Democratic Republic of the Congo). Here again, if 60 percent of annual U.S. auto sales were EVs by 2030 (7.8 million vehicles), those EVs (just in the U.S.) would consume about 78,000 tons of cobalt—half the world's supply.

LAWSUITS

As my publisher was preparing to submit this manuscript to the formatter who lives in Edmonton, Alberta, yet another headline made news. Between 2019 and 2021, the City of Edmonton purchased 60 electric buses from U.S. company Proterra. The purchase contract specified that the buses would perform in Edmonton's climate conditions with a warm temperature driving range of 328 kms and a cold temperature driving range of 268 kms. In its $82 million *Statement of Claim*, the City of Edmonton is claiming the bus range has been approximately 165 kms in the winter and, at best, 250 kms in warmer weather. Will this lawsuit yield any results? Not likely. Proterra declared Chapter 11 creditor protection in August 2023. Will there be more lawsuits against battery makers or electric vehicle makers? Count on it!

TESLA MISSES

While I expect many more negative headlines to emerge regarding electric vehicles, the news release from Tesla on April 2, 2024 was a big one that stopped stock market analysts in their tracks. For the first three months of 2024, the company fell short on its projected vehicle sales by over 62,000 units (a 13% miss and its biggest ever). The consumer is starting to speak.

- The theme of electric vehicles is volatile and becoming controversial. The pitfalls could be numerous for the consumer who buys an electric vehicle. As the old legal expression *Caveat Emptor* reminds us—*Buyer Beware*.

CONCLUSION

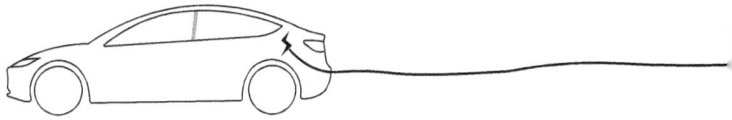

Collectively, the eight billion of us on the planet need to reduce energy consumption, reduce emissions, and pare back on our affluence. We cannot continue consuming more, acquiring more, and travelling more as the constructs of capitalism have so encouraged.

Rather than moderating our consumer habits, we have gratefully accepted from our elected officials what amounts to a get-out-of-jail-free card. We have been led to believe that our guilt arising from our overindulgence can be eased by shifting from internal combustion engines to electric vehicles. However, our elected officials have not single-handedly developed this panacea for our guilt. They have had help in promoting the electric vehicle theme from academia and from powerful groups like the IPCC, the World Economic Forum, and the Club of Rome.

But what these influencial groups have ignored is the fact that assembling a battery for an electric vehicle requires strategic metals.

These metals in the Earth's crust are non-renewable. When they are gone, they are gone for good; to the detriment of future generations who will inhabit the planet. We are already unsustainably mining these metals on a regular basis to manufacture the "stuff" that we think we need as we pursue affluence and material wealth. Extracting even more of these strategic metals to make batteries for electric vehicles only adds to the unsustainability argument.

Furthermore, it takes energy to make energy; the immutable Laws of Thermodynamics make this very clear. Where does the energy come from to charge a battery in an electric vehicle? To empower a vehicle, is it more energy efficient to burn coal at a power generating station to create electricity, or is it more energy efficient to extract petroleum from the ground and refine it into gasoline? The surprising answer is that battery-powered electric vehicles are no more energy efficient than vehicles powered by internal combustion engines. Instead of hydrocarbons, is it more efficient to use techniques like wind power or solar power to create electricity to charge vehicle batteries? The surprising answer is no. Green energy methods suffer from a lack of efficiency. Our elected officials are starting to impress upon us the benefits of generating electricity using nuclear reactors - which do not produce carbon dioxide emissions. But what do we do with all the nuclear waste that is accumulating on the planet? The bottom line is, when it comes to providing energy to empower vehicles, there is no free lunch.

Electric vehicles are nothing more than a knee-jerk, political reaction designed to divert our attention from the fact that we have no control over the Cycles of Nature that influence our planet. What is not being talked about by academia and elitist groups is the scientific fact that climate change is influenced by long cycles in the eccentricity of the Earth's orbit around the Sun, by cyclical variations in the angle of tilt of the Earth, and by cyclical variations in the rotational mechanics of the planet. Mankind cannot alter these cycles.

The electric vehicle theme is designed to ease our guilt, divert our attention, and make us think we are saving the planet. The promise that electric vehicles will save the planet is not only unsustainable, it is misguided and dangerous.

CHARGED!

NOTES

Introduction

Colten, J., Lindeberg, R. (2023) "Volta Trucks Files for Bankruptcy After Supply Chain Fails". [online] Available at: https://www.bloomberg.com/news/articles/2023-10-17/swedish-electric-vehicle-maker-volta-trucks-files-for-bankruptcy? Accessed: January 2024.

Daniel, W. (2024) "One of Wall Street's biggest Tesla bulls rips Elon Musk's 'train wreck' earnings call: 'We wrongly expected adults in the room'". [online] Available at: https://fortune.com/2024/01/25/tesla-stock-price-earnings-call-bulls-rip-musk. Accessed: January 2024.

Gomes, N., White, J. (2024) "Rental giant Hertz dumps EVs, including Teslas, for gas cars". [online] Available at: https://www.reuters.com/business/autos-transportation/hertz-sell-about-20000-evs-us-fleet-2024-01-11. Accessed: January 2024.

Shepardson, D. (2023) "Ford to temporarily cut one shift at Michigan EV plant". [online] Available at: https://www.reuters.com/business/autos-transportation/ford-temporarily-cut-one-shift-michigan-ev-plant-2023-10-13. Accessed: January 2024.

White, J. (2023) "Ford scales back Michigan battery plant, restarts construction". [online] Available at: https://www.reuters.com/business/autos-transportation/ford-scales-back-michigan-battery-plant-restarts-construction-2023-11-21. Accessed: January 2024.

Chapter 1

American Physical Society website (2023) This Month in Physics History. "March 20, 1800: Volta describes the Electric Battery". [online] Available at: https://www.aps.org/publications/apsnews/200603/history.cfm. Accessed: May 2023.

Continental Battery website (2023) "How It Works: The Step by Step of Lead-Acid Battery Recycling". [online] Available at https://www.continentalbattery.com/blog/how-it-works-the-step-by-step-of-lead-acid-battery-recycling. Accessed: May 2023.

Dummies website (2021) "Electrochemical Cells: The Danielle Cell". [online] Available at: https://www.dummies.com/article/academics-the-arts/science/chemistry/electrochemical-cells-the-daniell-cell-194214. Accessed: May 2023.

Heilbron, J.L. (1979) *Electricity in the 17th and 18th Centuries: A Study of Early Modern Physics*. University of California Press.

Hill, A. (2021) "What's the Lifespan of a lead acid battery?" Philadelphia Scientific website. [online] Available at: https://www.phlsci.com/news/articles/what-s-the-lifespan-of-a-lead-acid-battery. Accessed: May 2023.

Kurzweil, P. (2010) "Gaston Plante and his invention of the lead-acid battery – The genesis of the first practical rechargeable battery". *Journal of Power Sources*. Vol 195 (14) pp: 4424-4434. Accessed: May 2023.

National Museum of Scotland website (2023) "Bunsen Cell". [online] Available at: https://nms.scran.ac.uk/database/record.php?usi=000-190-004-730-C. Accessed: May 2023.

UPS Battery website (2014) "Ernst Waldemar Jungner and his Portable Batteries". [online] Available at: https://www.upsbatterycenter.com/blog/waldemar-jungner. Accessed: May 2023.

Wayback Machine website (2023) "The Kite Experiment". [online] Available at: http://franklinpapers.org/franklin/framedVolumes.jsp?vol=4&page=360a. Accessed: May 2023.

Chapter 2

Guarnieri, M. (2011) "When Cars Went Electric". *IEEE Industrial Electronics Magazine*. March 2011 and June 2011 issues.

Melosi, M. (2010) "The Automobile and the Environment in American History". [online] Available at: *http://www.autolife.umd.umich.edu/ Environment/E_Overview/E_Overview3.html*. Accessed: June 2023.

Super Cars website (2022) [online] "Early 1900s Cars". [online] Available at: https://www.supercars.net/blog/early-1900s-cars. Accessed: June 2023.

Truett, R. (2021) "A LONG HISTOR-E; Automakers have been tinkering with electric cars on and off for decades". *Automotive News*. Vol. 95, Issue 6982, p. 27.

Yergin, D, (1991) *The Prize*. Free Press, New York.

Chapter 3

OPEC website (2023) "Brief History". [online] Available at: https://www. opec.org/opec_web/en/about_us/24.htm. Accessed: June 2023.

Truett, R. (2021) "A LONG HISTOR-E; Automakers have been tinkering with electric cars on and off for decades". *Automotive News*. Vol (95), no. 6982.

US Dept of Energy website (2014) "The History of the Electric Car". [online] Available at https://www.energy.gov/articles/history-electric-car. Accessed: June 2023.

Chapter 4

Jha, A.R. (2012) "Batteries for Electric and Hybrid Vehicles". In: *Next-Generation Batteries and Fuel Cells for Commercial, Military, and Space Applications*. CRC Press, USA.

Moore, T. (2000) "Hybrid EVs: Making the Grid Connection". *EPRI Journal*, Vol. 25, Issue 4.

Snyder, K., et al (2009) "Hybrid Vehicle Battery Technology – The Transition From NiMH To Li-Ion". SAE International website. [online] Available at: https://www.sae.org/publications/technical-papers/ content/2009-01-1385. Accessed: June 2023.

Chapter 5

Battery University website (2023) "BU-205: Types of Lithium-ion". [online] Available at: https://batteryuniversity.com/article/bu-205-types-of-lithium-ion. Accessed: September, 2023.

Nobel Prize website (2023). "John B. Goodenough Biographical". [online] Available at: https://www.nobelprize.org/prizes/chemistry/2019/goodenough/biographical. Accessed: September, 2023.

Nobel Prize website (2023). "M. Stanley Whittingham Facts". [online] Available at: https://www.nobelprize.org/prizes/chemistry/2019/whittingham/facts. Accessed: September, 2023.

Nobel Prize website (2023). "Akira Yoshino Facts". [online] Available at : https://www.nobelprize.org/prizes/chemistry/2019/yoshino/facts. Accessed: September, 2023.

Sawai, T. (2020) "The invention of rechargeable batteries: An interview with Dr. Akira Yoshino, 2019 Nobel laureate". [online] Available at: https://www.wipo.int/wipo_magazine/en/2020/03/article_0004.html.

Chapter 6

Club of Rome website (2023) "History". [online] Available at: https://www.clubofrome.org/history. Accessed: September 2023.

Meadows, D., Randers, J., William, B. (1972) *The Limits to Growth*. Universe Books. USA.

Ritchie, H., Roser, M. (2019) "Land Use". [online] Available at: https://ourworldindata.org/land-use. Accessed: September 2023.

Chapter 7

Erlich, P.R., Holdren, J.P. (1971) "Impact of Population Growth". *Science*. vol 171, No. 3977, pp: 1212-1217.

Kaya, Y. (1989) "Impact of Carbon Dioxide Emission Control on GNP Growth". *IPSS Working Paper.* [online] Available at: https://www.scirp.org/reference/ReferencesPapers?ReferenceID=1021752. Accessed: January 2024.

Malthus, T. (1791) *An Essay on the Principle of Population as it Affects the Future Improvement of Society*, London, U.K.

Chapter 8

WEF website (2023) [online] Available at: https://www.weforum.org. Accessed: September 2023.

Chapter 9

Randers, J. (2008) "Global collapse—Fact or fiction?" *Futures.* (40) pp: 853-864.

Chapter 10

De Pryck, K, Hulme, M. (2023) *A Critical Assessment of the Intergovernmental Panel on Climate Change.* Cambridge University Press.

IPCC website (2023) [online] Available at: https://www.ipcc.ch/sr15/chapter/spm. Accessed: September 2023.

Chapter 11

Annual Reports website (2022) [online] "Form 10-K for General Motors". Available at: https://www.annualreports.com/HostedData/AnnualReports/PDF/NYSE_GM_2022.pdf. Accessed: September 2023.

Government of Canada website (2023) "Proposed regulated sales targets for zero-emission vehicles. Environment and Climate Change Canada". [online] Available at: https://www.canada.ca/en/environment-climate-change/news/2022/12/proposed-regulated-sales-targets-for-zero-emission-vehicles.html. Accessed: December 2023.

Government of Quebec website (2023) [online] Available at: https://www.legisquebec.gouv.qc.ca/en/pdf/cs/A-33.02.pdf. Accessed: December 2023.

Graham, S. (2021) "Chevy Bolt battery fix announced – is it enough?". [online] Available at: https://electrek.co/2021/04/29/chevy-bolt-battery-fix-announced. Accessed: February 2024.

ICCT website (2019) "International Council on Clean Transportation. Overview of global zero-emission vehicle mandate programs". [online] Available at: https://theicct.org/wp-content/uploads/2021/06/Zero-Emission-Vehicle-Mandate-Briefing-v2.pdf. Accessed: December 2023.

Chapter 12

Hammerich, K. (2008) "Canada's First Chiropractor". [online] Available at: https://www.cndoctor.ca/canadas-first-chiropractor-937. Accessed: September 2023.

Makhlouff, K. (2018) "Meet Gregory Kouri, The Lebanese Investor Behind Elon Musk & PayPal". [online] Available at: https://www.the961.com/gregory-kouri-lebanese-behind-elon-musk-and-paypal. Accessed: September 2023.

Chapter 13

NOAA website (2019) "The Carbon Cycle". [online] Available at: https://www.noaa.gov/education/resource-collections/climate/carbon-cycle. Accessed: September 2023.

Siegel, E. (2017) "How Much CO_2 Does A Single Volcano Emit?" [online] Available at: https://www.forbes.com/sites/startswithabang/2017/06/06/how-much-co2-does-a-single-volcano-emit/?sh=a2352b95cbf5. Accessed: October 2023.

The Guardian website (2023) [online] Available at: https://www.theguardian.com/world/2023/jun/27/canada-wildfires-released-record-breaking-carbon. Accessed: December 2023.

US Department of Energy website (2023) [online] Available at: https://www.energy.gov/science/doe-explainsthe-carbon-cycle. Accessed: September 2023.

Chapter 14

Hays, J., Imbrie, J., Shackleton, N. (1977). "Variations in the Earth's Orbit: Pacemaker of the Ice Ages". *Science.* pp:. 1121-32.

Koppelaar, R., Middelkoop, W. (2017) *The Tesla Revolution.* Amsterdam University Press.

NASA Observatory website (2000) "Milutin Milankovitch". [online] Available at: https://earthobservatory.nasa.gov/features/Milankovitch. Accessed: December 2023.

Plummer, C, McGeary D. (1982) *Physical Geology*, Wm. C. Brown Publishers, Iowa, USA.

Chapter 15

Battery University website (2023). "Learn About Batteries". [online] Available at: https://batteryuniversity.com/articles. Accessed: September, 2023.

Chapter 16

Alvarez, S. (2022) "Tesla 4680 cells compared with BYD Blade and CATL Qilin structural batteries". [online] Available at: https://www.teslarati.com/tesla-4680-vs-byd-blade-vs-catl-qilin-structural-batteries-video. Accessed: September 2023.

Anderson, M. (2011) "Pounds that Kill: The External Costs of Vehicle Weight". [online] Available at: https://www.nber.org/papers/w17170. Accessed: September 2023.

Aregay, T. (2020) "Tesla Achieves A Battery Breakthrough, Here Are The Manufacturing Advancements That Made It Possible". [online] Available at: https://www.torquenews.com/11826/tesla-achieves-battery-breakthrough-here-are-manufacturing-advancements-made-it-possible. Accessed: September 2023.

Aregay,T. (2021) "Tesla Signs A 5 Year Exclusive Deal With Jeff Dahn, Possibly The Best Battery Researcher In The World". [online] Available at: https://www.torquenews.com/11826/tesla-signs-5-year-exclusive-deal-jeff-dahn-possibly-best-battery-researcher-world. Accessed: September 2023.

Battery University website (2023). "BU-205: Types of Lithium-ion". [online] Available at:. https://batteryuniversity.com/article/bu-205-types-of-lithium-ion. Accessed: September 2023.

Bibienne, T., et al (2020) "From Mine to Mind and Mobiles: Society's Increasing Dependence on Lithium". *Elements*, Vol. 16, pp. 265–270.

Bomey, N. (2023) "EVs are much heavier than gas vehicles, and that's posing safety problems". [online] Available at: https://www.axios.com/2023/04/28/evs-weight-safety-problems. Accessed: September 2023.

Boyer,T. (2021) "Tesla 18650, 2170 and 4680 Battery Cell Comparison Basics". https://www.torquenews.com/14093/tesla-18650-2170-and-4680-battery-cell-comparison-basics. Accessed: September 2023.

Brunelist website (2022) "About NCMA, the Battery Chemistry Used in the Hummer EV". [online] Available at: https://www.brunelist.com/2022/05/02/about-ncma-the-battery-chemistry-used-in-the-hummer-ev. Accessed: February 2024.

Clean Technica website (2023) "Tesla 4680 battery production is trapped in production hell". [online] Available at: https://cleantechnica.com/2023/12/24/tesla-4680-battery-production-is-trapped-in-production-hell. Accessed: February 2024.

EPEC website (2024) "Prismatic and Pouch Battery Packs". [online] Available at: https://www.epectec.com/batteries/prismatic-pouch-packs.html. Accessed: February 2024.

EVBox website (2023) "What are electric car batteries made of?" [online] Available at: https://blog.evbox.com/what-are-ev-batteries-made-of? Accessed: September 2023.

History Computer website (2024) "4680 Battery: Everything You Need to Know About These New Cells". [online] Available at: https://history-computer.com/4680-battery-cells. Accessed: February 2024.

Imreh, A. (2021). "Blade – Cell To Pack LFP". [online] Available at: https://ibikes.wordpress.com/2021/01/03/blade-cell-to-pack-lfp. Accessed: September 2023.

Kane. M. (2021) "Check Electric Cars Listed By Weight Per Battery Capacity (kWh)". [online] Available at: https://insideevs.com/news/528346/ev-weight-per-battery-capacity. Accessed: September 2023.

Kothari, S. (2022) "BYD Blade Battery: Everything you should know". [online] Available at: https://topelectricsuv.com/news/byd/byd-blade-battery-update. Accessed: September 2023.

LG Energy website (2019) "This is Why NCM is the Preferable Cathode Material for Li-ion Batteries". [online] Available at: https://lghomebatteryblog.eu/en/this-is-why-ncm-is-the-preferable-cathode-material-for-li-ion-batteries. Accessed: February 2024.

Man. H. (2023) "What are LFP, NMC, NCA batteries in electric-cars?" [online] Available at: https://zecar.com/resources/what-are-lfp-nmc-nca-batteries-in-electric-cars. Accessed: September 2023.

Melancon, S. (2022) "Prismatic Cells vs Cylindrical Cells: What's the Difference?" [online] Available at: https://www.laserax.com/blog/prismatic-vs-cylindrical-cells. Accessed: September 2023.

My EV Review website (2023) "Electric Cars Weight Comparison Chart". [online] Available at: https://www.myevreview.com/comparison-chart/weight. Accessed: September 2023.

Push EVs website (2021) "NCM 712 by LG Chem: E66A and E78 battery cells in detail". [online] Available at: https://pushevs.com/2021/03/30/ncm-712-by-lg-chem-e66a-and-e78-battery-cells. Accessed: February 2024.

Reuters (2023) China's Gotion to set up $2 billion lithium battery plant in Illinois. [online] Available at: https://www.reuters.com/business/

autos-transportation/chinas-gotion-set-up-2-bln-lithium-battery-plant-illinois-2023-09-08. Accessed: February 2024.

Chapter 17

Abraham, K.M., (2020) "How Comparable Are Sodium-Ion Batteries to Lithium-Ion Counterparts". *ACS Energy Letters*, volume 5, number 11, pp: 3544-3547.

Flash Battery website (2023) "Solid-State Batteries: The New Frontier of Electrification". [online] Available at: https://www.flashbattery.tech/en/how-solid-state-batteries-work. Accessed: September 2023.

Chapter 18

Butler, C. (2023) "Canada is paying 'an enormous price' for the Volkswagen battery plant. Is it worth it?" [online] Available at: https://www.cbc.ca/news/canada/london/st-thomas-ev-battery-plant-volkswagen-subsidy-1.6876156. Accessed: September 2023.

GM Cami website (2023) "CAMI Assembly". [online] Available at: https://www.gmcamiassembly.ca/en/home/company/canada/cami.html. Accessed: September 2023.

Gohlke, D., Zhou, Y., Wu, X., Courtney, C. (2022) "Assessment of Light-Duty Plug-in Electric Vehicles in the United States, 2010–2021". [online] Available at: https://publications.anl.gov/anlpubs/2022/11/178584.pdf. Accessed: September 2023.

Kane, M. (2023) "Expansion Of EV Battery Manufacturing Capacity In North America Amazes". [online] Available at: https://insideevs.com/news/654889/ev-battery-manufacturing-capacity-north-america-2030. Accessed: September 2023.

Noble, B. (2023) "Stellantis, Samsung SDI will team to build a second EV battery plant in United States". [online] Available at: https://www.detroitnews.com/story/business/autos/chrysler/2023/07/24/stellantis-samsung-sdi-second-us-ev-battery-plant/70454364007. Accessed: September 2023.

Chapter 19

Ahead of The Herd website (2019). "Is China Locking Up Indonesian Nickel?"[online] Available at: https://aheadoftheherd.com/is-china-locking-up-indonesian-nickel. Accessed: September 2023.

Ahead of The Herd website (2019). "The Future of Canada's Nickel Supply Isn't Indonesia". [online] Available at: https://aheadoftheherd.com/the-future-of-canadas-nickel-supply-isnt-indonesia. Accessed: September 2023.

Ahead of The Herd website (2019). "The Case for Low Grade Sulfide Nickel Deposits".[online] Available at: https://aheadoftheherd.com/the-case-for-low-grade-sulfide-nickel-deposits. Accessed: September 2023.

Alexeev, S. et al (2020) "Brines of the Siberian platform (Russia): Geochemistry and processing prospects". *Applied Geochemistry* 117 (2020) 104588.

Barrera, P. (2017) "Manganese Reserves by Country". [online] Available at: https://investingnews.com/daily/resource-investing/battery-metals-investing/manganese-investing/manganese-reserves. Accessed: September 2023.

Bi, Y, (2016) "Stability of Li2CO3 in cathode of lithium ion battery and its influence on electrochemical performance". *RSC Advances*, Vol 6, 19233.

Bibienne, T. et al (2020) "From Mine to Mind and Mobiles: Society's Increasing Dependence on Lithium", *Elements*, vol. 16., pp: 265-270.

Bogossian, J. (2021) "Brine Lithium Deposits". [online] Available at: https://www.geologyforinvestors.com/brine-lithium-deposits. Accessed: September 2023.

Canadian Manganese website (2023) "HPMSM Processing 101". [online] Available at: https://canadianmanganese.com/hpmsm/hpmsm-processing-101. Accessed: September 2023.

Cannon, W., Kimball, B., Corathers, L. (2017) "Critical Mineral Resources of the United States—Economic and Environmental Geology and Prospects for Future Supply". U.S. Geological Survey. [online] Available at: https://pubs.usgs.gov/pp/1802/l/pp1802l.pdf. Accessed: September 2023.

Cannon, W. et al (2017) "Manganese". In: *Critical Mineral Resources of the United States—Economic and Environmental Geology and Prospects for Future Supply*. Eds: Klaus J. Schulz, et al. US Geological Survey.

Chakhmouradian, A. R. & Wall, F. (2012). "Rare earth elements: minerals, mines, magnets, (and more)". *Elements*, 8(5), 333-340.

Dushyantha, N. et al (2020) "The story of rare earth elements (REEs): Occurrences, global distribution, genesis, geology, mineralogy and global production". *Ore Geology Reviews*. Volume 122, 103521.

Guiomar ,C. (2016) "Decreasing Ore Grades in Global Metallic Mining: A Theoretical Issue or a Global Reality?" Resources, 5(4), 36; https://doi.org/10.3390/resources5040036.

Horn, S. et al (2021) "Cobalt resources in Europe and the potential for new discoveries". Ore Geology Reviews, vol. 130. pp:1-25. Irle, R. (2022) "Global EV Sales for 2022". [online] Available at: https://www.ev-volumes.com. Accessed: September 2023.

Jamasmie, C. (2022) [online] "Electric vehicles surpass phones as top driver of cobalt demand". Available at: https://www.mining.com/electric-vehicles-surpass-phones-as-top-driver-of-cobalt-demand. Accessed: September 2023.

Lithium Americas website (2018) "Technical Report on the PreFeasibility Study for the Thacker Pass Project, Humboldt County, Nevada, USA 2018". [online] Available at: https://www.lithiumamericas.com/_resources/pdf/investors/technical-reports/thacker-pass/Technical-Report-Thacker-Pass.pdf. Accessed: September 2023.

Liu, W. (2019) "Spatiotemporal patterns of lithium mining and environmental degradation in the Atacama Salt Flat, Chile". *International Journal of Applied Earth Observation and Geoinformation*, Volume 80, pp: 145-156.

Lorca, M. (2022) "Mining indigenous territories: Consensus, tensions and ambivalences in the Salar de Atacama". *The Extractive Industries and Society*, Volume 9, 101047.

Meng, J. (2023) "An Overview of World Nickel Resources". [online] Available at: *https://stainless-steel-world.net/an-overview-of-world-nickel-resources*. Accessed: September 2023.

MiningIR website (2019) "Lithium: Outlook to 2028. Roskill Reports on Metals and Minerals 16th Edition". [online] Available at: https://miningir.com/roskill-releases-updated-lithium-outlook-believes-weak-prices-to-claim-more-projects-before-recovery. Accessed: September 2023.

Mining-Technology website (2023) "Bolivia's YLB signs lithium agreements with Russian and Chinese companies". [online] Available at: https://www.mining-technology.com/news/ylb-lithium-russian-chinese. Accessed: September 2023.

Mohr, S. et al (2012) "Lithium Resources and Production: Critical Assessment and Global Projections". *Minerals*, Vol. 2., pp: 65-84.

Morissette, C. (2012) "The Impact of Geological Environment on the Lithium Concentration and Structural Composition of Hectorite Clays". M.Sc. Thesis, U of Nevada, Reno, USA. [online] Available at: https://scholarworks.unr.edu/bitstream/handle/11714/3586/LamyMorissette_unr_0139M_10965.pdf?sequence=1&isAllowed=y. Accessed: September 2023.

Nangoy, F., Ungku, F. (2021) "Exclusive: Facing green pressure, Indonesia halts deep-sea mining disposal". [online] Available at: https://ca.finance.yahoo.com/news/exclusive-facing-green-pressure-indonesia-085043133.html. Accessed: September 2023.

Narayanan, A. (2023) " China EV Sales Improve For Nio And Xpeng, While BYD And Li Auto Boom And Hit Milestones". [online] Available at: https://www.investors.com/news/china-ev-sales-june-q2-li-auto-nio-xpeng-byd-tesla/. Accessed: September 2023.

Roberts. S., Gunn, G. (2014) "Cobalt". In: *Critical Metals Handbook*, First Edition. Ed. Gus Gunn. John Wiley & Sons Ltd.

Romero, V.C. (2021) "Electrochemical extraction of lithium by ion insertion from natural brine using a flow-by reactor: Possibilities and limitations". *Electrochemistry Communications*. Volume 125, 106980.

Smith, C.G. (2001) "Always the bridesmaid, never the bride: cobalt geology and resources". *Applied Earth Science*. 110:2, pp: 75-80.

Sprott Money website (2022) "The COMEX Silver Vaults". [online] Available at: https://www.sprottmoney.com/blog/The-COMEX-Silver-Vaults-November-04-2022. Accessed: September 2023.

USGS website (2022) "USGS Mineral Commodity Summaries 2022". [online] Available at: https://pubs.usgs.gov/periodicals/mcs2022/mcs2022. pdf. Accessed: September 2023.

USGS website (2020) "Lithium Occurrences and Processing Facilities of Argentina, and Salars of the Lithium Triangle, Central South America". [online] Available at: https://www.usgs.gov/data/lithium-occurrences-and-processing-facilities-argentina-and-salars-lithium-triangle-central. Accessed: September 2020.

USGS website (2023) [online] "Silver" [online] Available at: https://pubs. usgs.gov/periodicals/mcs2023/mcs2023-silver.pdf. Accessed: September 2023.

Vasyukova, O.V., Williams-Jones, A.E. (2022) "Constraints on the Genesis of Cobalt Deposits: Part II. Applications to Natural Systems". *Economic Geology*, v. 117, no. 3, pp. 529–544.

Vera.M. et al (2023) "Environmental impact of direct lithium extraction from brines". *Nature Reviews Earth & Environment*. Vol. 4, pages 149–165.

Wei,W. et al (2020) "Energy Consumption and Greenhouse Gas Emissions of Nickel Products". *Energies, 13*(21), 5664.

Yusifova, N.V. et al (2019) "Development of a Method for Cobalt Recovery from Cobalt Ore". *Russian Metallurgy (Metally)*, Vol. 2019, No. 3, pp. 204–209.

Zhang, T. et l (2021) "Cradle-to-gate life cycle assessment of cobalt sulfate production derived from a nickel–copper–mine in China". *The International Journal of Life Cycle Assessment.* Vol 26, pp: 1198-1210.

Chapter 20

Atkins, P.W. (1982) *Physical Chemistry.* Oxford University Press. U.K.

Chapter 21

Bodini, N, et al (2021) "Wind plants can impact long term local atmospheric conditions". *Scientific Reports*, volume 11, Article number: 22939.

EPA Website (2023) "Renewable Energy Fact Sheet: Wind Turbines". [online] Available at: https://www.epa.gov/sites/default/files/2019-08/documents/wind_turbines_fact_sheet_p100il8k.pdf. Accessed: September 2023.

Chapter 22

Chint Global website (2023) "How Much Power Loss in Transmission Lines?". [online] Available at: https://chintglobal.com/blog/how-much-power-loss-in-transmissionlines. Accessed: September, 2023.

Concawe website (2012) "EU refinery energy systems and efficiency". [online] Available at: https://www.concawe.eu/wp-content/uploads/2017/01/rpt_12-03-2012-01520-01-e.pdf. Accessed: September 2023.

EIA website (2023) "What is the heat content of U.S. coal?". [online] Available at: https://www.eia.gov/tools/faqs/faq.php?id=72&t=2. Accessed: September 2023.

EPA website (2023) "New Coal-Fired Power Plant Performance and Cost Estimates". [online] Available at: https://www.epa.gov/sites/default/files/2015-08/documents/coalperform.pdf. Accessed: September 2023.

Hong, B., Slatick, E. (1994) "Carbon Dioxide Emission Factors for Coal". Energy Information Administration, Quarterly Coal Report, January-April 1994. [online] Available at: https://www.eia.gov/coal/production/quarterly/co2_article/co2.html. Accessed: September 2023.

ICCT website (2010) "Carbon Intensity of Crude Oil in Europe. A report by Energy-Redefined LLC for the International Council on Clean Transportation". [online] Available at: https://theicct.org/sites/default/files/ICCT_crudeoil_Eur_Dec2010_sum.pdf. Accessed: September 2023.

Natural Resources Canada (2023) [online] Available at: 'Fuel Efficient Technologies". https://natural-resources.canada.ca/sites/nrcan/files/oee/pdf/transportation/fuel-efficient-technologies/autosmart_factsheet_9_e.pdf. Accessed: September 2023.

Raghuvanshi, S. et al (2006) "Carbon dioxide emissions from coal based power generation in India". *Energy Conversion and Management*, 47, pp: 427–441.

US EPA website (2023) "US EPA emissions". [online] Available at: https://nepis.epa.gov/Exe/ZyNET.exe. Accessed: September 2023.

Chapter 23

Boretti, A, Banik, B. (2021) "Advances in Hydrogen Production from Natural Gas Reforming". *Advanced Energy and Sustainability Research.* 2: 2100097. [online] Available at: https://doi.org/10.1002/aesr.202100097. Accessed: December 2023.

Collins. L. (2023) "Plug Power to invest $365m in Korean gigafactory as part of previously announced joint venture". [online] Available at: https://www.hydrogeninsight.com/electrolysers/plug-power-to-invest-

365m-in-korean-gigafactory-as-part-of-previously-announced-joint-venture/2-1-1444337. Accessed: December 2023.

Collins. L. (2022) "Hydrogen is starting to look like an economic bubble — and here's why". [online] Available at: https://www.hydrogeninsight.com/analysis/liebreich-hydrogen-is-starting-to-look-like-an-economic-bubble-and-here-s-why. Accessed: December 2023.

Collins, L. (2023) "Norwegian electrolyser manufacturer Nel has announced that it will build a long-planned 4GW gigafactory in the US state of Michigan, at a cost of up to $400m". [online] Available at: https://www.hydrogeninsight.com/electrolysers/nel-announces-new-4gw-hydrogen-electrolyser-gigafactory-in-michigan-costing-up-to-400m. Accessed: December 2023.

Collins, L. (2023) "First e-fuel made from green hydrogen and CO_2 is 100 times more expensive than petrol, but costs should plummet". [online] Available at: https://www.hydrogeninsight.com/transport/first-e-fuel-made-from-green-hydrogen-and-co2-is-100-times-more-expensive-than-petrol-but-costs-should-plummet. Accessed: December 2023.

Jha, A.R. (2012) "Batteries for Electric and Hybrid Vehicles". In: *Next-Generation Batteries and Fuel Cells for Commercial, Military, and Space Applications.* CRC Press, New York, USA.

Martin, P. (2023) "Will rising platinum and iridium prices restrict the growth of PEM hydrogen electrolysers and fuel cells?" [online] Available at: https://www.hydrogeninsight.com/analysis/analysis-will-rising-platinum-and-iridium-prices-restrict-the-growth-of-pem-hydrogen-electrolysers-and-fuel-cells-/2-1-1460113. Accessed: December 2023.

Pilatowsky, I. et al (2011) "Thermodynamics of Fuel Cells". In: *Cogeneration Fuel Cell-sorbtion Air Conditioning Systems.* Springer, London, U.K.

US Dept of Energy website (2023) "Hydrogen Laws and Incentives in California". [online] Available at: https://afdc.energy.gov/fuels/laws/HY. Accessed: December 2023.

Chapter 24

AZO Materials website (2023) "Advanced Materials in Jet Engines". [online] Available at: https://www.azom.com/article.aspx?ArticleID=90. Accessed: September 2023.

Clemente, J. (2016) "U.S. Natural Gas Electricity Efficiency is Always Improving". [online] Available at: https://www.forbes.com/sites/judeclemente/2016/04/10/u-s-natural-gas-electricity-efficiency-continues-to-improve/?sh=174f6c9e35a4. Accessed: September 2023.

EIA website (2023) "Natural gas explained". [online] Available at: https://www.eia.gov/energyexplained/natural-gas/how-much-gas-is-left.php. Accessed: September 2023.

Energy.gov website (2023) "How Gas Turbine Power Plants Work". [online] Available at: https://www.energy.gov/fecm/how-gas-turbine-power-plants-work. Accessed: September 2023.

Guerrero. M. (2023) "Turbine Efficiency Formula: Gas Turbine Efficiency Calculations". [online] Available at: https://www.araner.com/blog/gas-turbine-efficiency-formula. Accessed: September 2023.

Chapter 25

American History website (2023) "Fuel Cells". [online] Available at: https://americanhistory.si.edu/fuelcells/pem/pemmain.htm. Accessed: December 2023.

Larson, L. (2020) "Nuclear Waste Storage Sites in the United States". [online] Available at: https://sgp.fas.org/crs/nuke/IF11201.pdf. Accessed: December 2023.

Macfarlane, A., Ewing, R. (2023) "Nuclear Waste Is Piling Up. Does the U.S. Have a Plan?". [online] Available at: https://www.scientificamerican.com/article/nuclear-waste-is-piling-up-does-the-u-s-have-a-plan. Accessed: December 2023.

Vattenfall website (2023) "Finland to open the world's first final repository for spent nuclear fuel". [online] Available at: https://group.vattenfall.com/press-and-media/newsroom/2023/finland-to-open-the-worlds-first-final-repository-for-spent-nuclear-fuel. Accessed: February 2024.

World Nuclear website (2023) "Supply of Uranium". [online] Available at: https://www.world-nuclear.org/information-library/nuclear-fuel-cycle/uranium-resources/supply-of-uranium. Accessed: December 2023.

Chapter 26

Shockley, W., Queisser, H. (1961) "Detailed Balance Limit of Efficiency of p-n Junction Solar Cells". *Journal of Applied Physics*, Volume 32, pp. 510-519.

Chapter 27

Durante, D. Sneller, T. (2009) "The Net Energy Balance of Ethanol Production". [online] Available at: http://large.stanford.edu/courses/2014/ph240/dikeou1/docs/net_energy_balance2009.pdf. Accessed: December 2023.

Modern Power Systems website (2023) "Living on Alcohol and Anti-freeze" [online] Available at: https://www.modernpowersystems.com/features/featureliving-on-alcohol-and-anti-freeze. Accessed: December 2023.

Schmer, M. et al (2008) "Net energy of cellulosic ethanol from switchgrass". *PNAS*, 105 (2) 464-469.

Chapter 28

Angueira, G. (2023) "The US Doesn't Have a Law Mandating EV Battery Recycling". [online] Available at: https://grist.org/technology/the-u-s-doesnt-have-a-law-mandating-ev-battery-recycling-should-it. Accessed: September 2023.

Baum, Z. et al (2022) "Lithium-Ion Battery Recycling Overview of Techniques and Trends". *ACS Energy Letters.* 7 (2), pp: 712-719.

Holzer, A. et al (2021) "A Novel Pyrometallurgical Recycling Process for Lithium-Ion Batteries and Its Application to the Recycling of LCO and LFP". *Metals* 11(1), p.149.

"Hydrometallurgical process for recovery of metal values from spent lithium-ion secondary batteries". *Hydrometallurgy*, Volume 47, Issues 2–3, pp: 259-271.

Notter, D. (2010) "Contribution of Li-Ion Batteries to the Environmental Impact of Electric Vehicles". *Environmental Science and Technology*, 44, pp: 6550–6556.

Peeters, N. (2022) "Recovery of cobalt from lithium-ion battery cathode material by combining solvo-leaching and solvent extraction". *Green Chemistry*. (24), pp:2839-2852.

Zhang, P., Yokoyama, T., Itabashi, O., Suzuki, T., Inoue, K. (1998)

Chapter 29

Erwin, B. (2020) "How Does The EPA Calculate Electric Car Range?" [online] Available at: https://cleantechnica.com/2020/08/18/how-does-epa-calculate-electric-car-range. Accessed: December 2023.

Holderith, P. (2022) "General Motors Will Lose Money on Its Electric Cars Until 2025". [online] Available at: https://www.thedrive.com/news/general-motors-will-lose-money-on-its-electric-cars-until-2025-report. Accessed: December 2023.

Khanna, N.(2023) "Does DC Fast Charging Really Reduce Your EV's Battery Capacity?" [online] Available at: https://www.makeuseof.com/dc-fast-charging-bad-for-evs. Accessed: December 2023.

Nimmo, J. (2023) "EVs Cost Twice as Much to Insure as Fuel-Burning Cars in UK". [online] Available at: https://www.bnnbloomberg.ca/evs-cost-twice-as-much-to-insure-as-fuel-burning-cars-in-uk-1.2025749. Accessed: December 2023.

Olinga, L. (2023) "Ford Loses Nearly $60,000 for Every Electric Vehicle Sold". [online] Available at: https://www.thestreet.com/technology/ford-loses-nearly-60000-for-every-electric-vehicle-sold. Accessed: December 2023.

Pannone, G.,VanderWerp, D. (2023) "Comparison of On-Road Highway Fuel Economy and All-Electric Range to Label Values: Are the Current Label Procedures Appropriate for Battery Electric Vehicles?". SAE Technical Paper 2023-01-0349. [online] Available at: https://doi.org/10.4271/2023-01-0349, Accessed: December 2023.

Priddle, A. (2022) "Stellantis CEO Carlos Tavares Breaks Down the True Cost of All Its New EVs". [online] Available at: https://www.motortrend.com/news/stellantis-jeep-dodge-2021-earnings-update-ev-cost. Accessed: December 2023.

Stecklow, S., Shirouzu, N., (2023) "Tesla created secret team to suppress thousands of driving range complaints". [online] Available at: https://www.reuters.com/investigates/special-report/tesla-batteries-range. Accessed: December 2023.

Transport Policy website (2023) "Worldwide Light Vehicles Test Procedure". [online] Available at: https://www.transportpolicy.net/standard/international-light-duty-worldwide-harmonized-light-vehicles-test-procedure-wltp. Accessed: December 2023.

US News website (2024) "Biden Administration Will Investigate National Security Risks Posed by Chinese-Made 'Smart Cars' ". [online] Available at: https://www.usnews.com/news/business/articles/2024-02-29/biden-administration-to-investigate-national-security-risks-posed-by-chinese-made-smart-cars. Accessed: February 2024.

Conclusion

Bloomberg website (2024) "GM Goes Back to the Future with Plans for Plug-In Hybrids". [online] Available at: https://www.bloomberg.com/news/articles/2024-01-30/gm-goes-back-to-the-future-with-plans-for-plug-in-hybrids?embedded-checkout=true. Accessed: February 2024.

Mulcahy, K. (2024) "We were blindsided': Edmonton seeking $82M in damages from U.S. company over electric buses". [online] Available at: https://edmonton.ctvnews.ca/we-were-blindsided-edmonton-seeking-82m-in-damages-from-u-s-company-over-electric-buses. Accessed: March 2024.

ABOUT THE AUTHOR

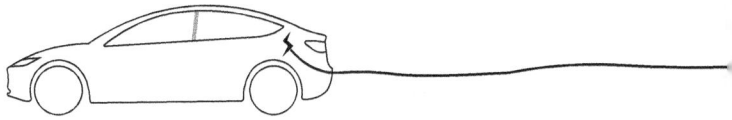

Malcolm Bucholtz holds an Engineering degree from Queen's University, and both an MBA and a M.Sc. degree from Heriot Watt University in Edinburgh, Scotland.

Malcolm is a researcher and author of more than twenty books on geopolitics, science, the financial markets, and future trends.

He lives in the small farming community of Mossbank, Saskatchewan, Canada.

There are eight billion of us on the planet and most of us strive for lifestyles of affluence, consumption, and mobility. We buy what we want and we go where we want to go.

These capitalistic-driven, hedonistic desires consume non-renewable energy and non-renewable mineral resources. If left unchecked, our consumption and mobility desires will deprive future generations of sufficient quantities of these resources.

This is not news.

However, rather than moderating our consumption habits, we have gratefully accepted a get-out-of-jail-free card. We have been led to believe that our overindulgence can be countered by making a shift to electric vehicles. We have also been led to believe that adopting electric vehicles is a solution to the climate change crisis. Guided by academia and elitist groups such as the IPCC, the World Economic Forum, and the Club of Rome, our elected officials have embraced this panacea for our collective guilt.

Unfortunately, the primary cause of climate change is the cyclical patterns in the Earth's orbital eccentricity, its tilt axis, and its rotational spin. Mankind is not responsible for creating these cycles; nor can mankind change them.

The electric vehicle strategy is not only misguided—it is dangerous. As this book explains, the electric car strategy not only is gobbling up our resources, it ignores the Laws of Thermodynamics—it takes energy to make energy. The energy to charge an electric vehicle battery has to come from somewhere. There is no free lunch.

Malcolm Bucholtz holds an Engineering degree from Queen's University, and both an MBA and a M.Sc. degree from Heriot Watt University (Edinburgh, Scotland). Malcolm is a researcher and author of more than 20 books on geopolitics, science, and the financial markets. Malcolm lives in the small farming community of Mossbank, Saskatchewan, Canada.

www.ingramcontent.com/pod-product-compliance
Lightning Source LLC
Chambersburg PA
CBHW071340210326
41597CB00015B/1520